SUSTAINABLE MARINE RESOURCE UTILIZATION IN CHINA

SUSTAINABLE MARINE RESOURCE UTILIZATION IN CHINA

A Comprehensive Evaluation

MALIN SONG

*School of Statistics and Applied Mathematics,
Anhui University of Finance and Economics,
Bengbu, P.R. China*

XIONGFENG PAN

*School of Economics and Management,
Dalian University of Technology,
Dalian, P.R. China*

XIANYOU PAN

*School of Economics and Management,
Dalian University of Technology,
Dalian, P.R. China*

ELSEVIER

Elsevier
Radarweg 29, PO Box 211, 1000 AE Amsterdam, Netherlands
The Boulevard, Langford Lane, Kidlington, Oxford OX5 1GB, United Kingdom
50 Hampshire Street, 5th Floor, Cambridge, MA 02139, United States

Notices
Knowledge and best practice in this field are constantly changing. As new research and experience broaden our understanding, changes in research methods, professional practices, or medical treatment may become necessary.

Practitioners and researchers must always rely on their own experience and knowledge in evaluating and using any information, methods, compounds, or experiments described herein. In using such information or methods they should be mindful of their own safety and the safety of others, including parties for whom they have a professional responsibility.

To the fullest extent of the law, neither the Publisher nor the authors, contributors, or editors, assume any liability for any injury and/or damage to persons or property as a matter of products liability, negligence or otherwise, or from any use or operation of any methods, products, instructions, or ideas contained in the material herein.

British Library Cataloguing-in-Publication Data
A catalogue record for this book is available from the British Library

Library of Congress Cataloging-in-Publication Data
A catalog record for this book is available from the Library of Congress

ISBN: 978-0-12-819911-4

For Information on all Elsevier publications
visit our website at https://www.elsevier.com/books-and-journals

Publisher: Candice Janco
Acquisition Editor: Glyn Jones
Editorial Project Manager: Andrea Dulberger
Production Project Manager: Sruthi Satheesh
Cover Designer: Miles Hitchen

Typeset by MPS Limited, Chennai, India

Working together
to grow libraries in
developing countries

www.elsevier.com • www.bookaid.org

Contents

Preface

As the bearing capacity of the land becomes increasingly limited, people from all walks of life are paying more attention to seeking new development space and sources of growth. Marine resources are the "second frontier" on which human beings rely for survival. Marine countries have successively formulated and adjusted their marine strategies. Two representative plans of the United States are "An Ocean Blueprint for the 21st Century" and "Ocean Action Plan"; in 2001 and 2015, Russia issued two versions of the "Maritime Doctrine of the Russian Federation"; Japan updates its "Basic Ocean Plan" every 5 years and the one published in 2012 formulated the framework for its maritime strategy from 2013 to 2017; in 2012, China developed the strategy of building itself into a maritime power, exploring the ocean, valuing the ocean, and making strategic plans regarding the ocean with unprecedented enthusiasm. The competition among major countries around the world to put forward long-term plans for their maritime strategies is a reflection of the rising influence of the ocean on a country's security and development interests. In the new era of international competition, it has become an important task for all governments to gain an advantage in maritime competition. However, with the acceleration of development of marine economy, consequent problems such as marine resource constraint, environmental pollution, and ecological security have become increasingly serious. People are gradually realizing that the traditional extensive mode that focused on marine resources development can no longer meet the needs of the current development situation, and it is an urgent problem to realize the transformation of marine resources utilization to the intensive growth mode that is guided by science and technology effectively.

As a large coastal country, China's marine economy has driven the rapid economic development in its eastern region, providing a powerful condition for this region to realize economic opening up. The land area of coastal provinces and cities accounts for about 13% of the total national area, but almost 40% of the national population are concentrated in such areas, which contribute more than 60% of GDP. Promoting the development and utilization of marine resources will become an urgent need to ease domestic resource constraints and expand land development space under the new situation. This book aims to break through theoretical research in accordance with the procedure of condensing key issues, elaborating basic theories, analyzing realistic conditions, evaluating utilization efficiencies, evaluating utilization impacts, explaining management effects, and designing policy systems to determine feasible countermeasures that can effectively support the decision-making practice of sustainable utilization and management evaluation of China's marine resources. First, core problems of sustainable exploitation, utilization, and management of marine resources under the realistic background of building a maritime power are condensed. Second, the sustainable utilization of marine resources, development of marine economy, and management of marine resources are theoretically analyzed by referring to relevant international theoretical achievements. Third, the marine ecological carrying capacity of China is evaluated and the basic conditions for the sustainable utilization of marine resources in China's coastal areas are analyzed. Fourth, economic, environmental, and comprehensive benefits of China's marine resources utilization are researched, and the realistic basis for the sustainable utilization of marine resources in China's coastal areas is clarified. Fifth, the coordinating and coupling relationship among China's marine resources utilization, economy, and environment is analyzed to grasp the social influences of sustainable utilization of marine resources. Finally, China's marine resources management level is evaluated and quality differences in sustainable utilization and management of such resources are explored.

The research of this book explores the basic problems and practical countermeasures of sustainable utilization of marine resources. It differs from other similar books in the following three aspects. First, it constructs the basic analytical framework of sustainable development evaluation of marine resources from the perspective of the trinity of marine resources, environment, and economy. Second, the rationality analysis and improvement of relevant evaluation methods are conducted to ensure that the evaluation results are scientific and effective. Third, under the background of the marine power strategy, basic conditions, basic status quo, and impacts on economic and social development of sustainable utilization of China's marine resources are comprehensively evaluated. Therefore this book is not only helpful for readers to have an in-depth understanding of the basic situation of sustainable development and utilization of China's marine resources under the background of the marine power strategy, but can also provide theoretical and methodological reference for readers to conduct sustainable utilization evaluation of other countries' marine resources.

I would like to express my heartfelt thanks to Ms. Mengna Li, Ms. Bowei Ai, and Mr. Yang Ming for their help in data collection, model calculation, and other aspects during the writing of this book. In addition, I would also like to thank all the experts and scholars who have put forward valuable opinions and suggestions during the review of this book. Finally, I would like to thank Elsevier Press for its support and help in editing and publishing this book.

This book is a preliminary exploration of evaluating China's sustainable utilization of marine resources. Due to time constraints and the authors' limited ability, there are inevitably omissions in the book. Criticism and comments are sincerely invited from every reader and colleague.

Malin Song, Xiongfeng Pan and Xianyou Pan
November 2018

CHAPTER ONE

An overview of sustainable marine resource utilization

1.1 Introduction

The intensive exploitation and utilization of marine resources in all the countries of the world indicates the significance of these resources for human survival. This impact is evident in the case of China's coastal economy. Specifically, the country's marine economy has contributed toward the rapid economic development of the eastern region, thereby providing a suitable environment for the region to achieve economic opening. In 2017 China's marine economic output reached CNY 7.8 trillion, with a year-on-year growth rate of nearly 7%. This implies the need to develop and utilize marine resources to ease domestic resource constraints and expand land development space under the new normal economy. However, with the acceleration and increase of the scale of marine economic development, problems such as the constraints of marine resources, environmental pollution, and ecological security are becoming increasingly serious. Gradually, people realize that the traditional extensive mode dominated by the exploitation of marine resources can no longer meet the needs of the current development situation. How to effectively realize the intensive growth of marine resource utilization is an urgent problem that needs to be solved. Today, the key to the sustainable utilization of marine resources lies in the appropriate development and utilization of marine resources, based on advanced technologies and materials, and ensuring balanced exploitation of the marine ecosystem. Concerning China's actual situation, the unreasonable spatial planning for marine development and the constraints associated with the technological level of marine resource utilization frequently lead to problems that affect the sustainable development of the marine economy, such as resource wastage and environmental pollution. Meanwhile, with the expansion of population size and rising economic pressure in coastal areas, the challenge of realizing sustainable development of the marine economy is becoming increasingly acute. Therefore although a necessary

Sustainable Marine Resource Utilization in China.
DOI: https://doi.org/10.1016/B978-0-12-819911-4.00001-1

condition for the large-scale development of China's marine economy in the future, it has become challenging to systematically implement the idea of sustainable development in the development and utilization of marine resources to realize the harmonious development of marine resources, environment, and the economy. This chapter starts with summarizing the concept of the sustainable utilization of marine resources and related theories, and the main practical experience of international marine resource management is elaborated to provide a theoretical reference for the follow-up study on the sustainable utilization of marine resources. Subsequently, the key issues and research contents to be solved in this book are put forward, and the corresponding solutions and methods are designed under a comprehensive and unique approach.

1.2 Overview of theories on sustainable marine resource utilization

1.2.1 Concept of sustainable marine resource utilization

The sustainable development of marine resources refers to the mutual promotion of economy, ecology, and society to achieve sustainable development. Sustainable development is a development model that considers the environment when achieving balanced growth with efficiency and fairness. In addition, it focuses on the short- and long-term local and overall interests while contributing toward the socioeconomic systems and individuals' lifestyles. The sustainable development theory emphasizes maintaining the integrity and sustainability of development. In this case, integrity refers to achieving the developmental goals by considering both local and overall interests. In other words, the overall interests (interests of all the stakeholders) should not be sacrificed for regional interests during development; various systems should coordinate and promote each other to achieve common progress.

First, sustainable development emphasizes the coordinated development between resource utilization and economic growth. The countries and regions recording a relatively backward economic development must change the traditional development mode, which neglects environmental protection and resource saving, and emphasize cleaner production and civilized consumption to achieve sustainable economic growth. This can be achieved through a combination of rational resource utilization and environmental protection. The level of the contemporary economy should be improved while ensuring the goal of sustainable economic growth in the future.

Second, the theory of sustainable development emphasizes moderate development, taking into account the environmental carrying capacity while using natural resources. In the process of development, we should fully consider the scarcity and nonrenewability of resources; ensure rational development, utilization, and conservation of resources; improve the cleaner production and environmental self-protection capacities; and improve the utilization rate of environmental resources, to establish a stable ecosystem and realize the development goals.

Finally, the main goal of the sustainable development theory is to improve the quality of human life. In other words, sustainable development emphasizes that human society should reduce the gap of wealth and unemployment rate by controlling population growth rate and by improving and regulating social distribution to ultimately establish a geo-social environment, in which people can live and work in peace and contentment and have a high quality of life.

1.2.2 Connotation of marine resource management

1.2.2.1 Origin of comprehensive management of marine resources

Since the 1930s, the idea of comprehensive management of marine resources has been gradually put forward with the development and utilization of marine resources. In the beginning, some scholars suggested that comprehensive management should be adopted to coordinate and manage the development between the outer margin space of the continental shelf and marine resources. In the late 1970s, owing to marine resource constraints and marine environment deterioration, the idea of comprehensive marine management was further emphasized. In 1982 the United Nations adopted the United Nations Convention on the Law of the Sea, which laid the foundation for the follow-up comprehensive marine management. In 1989 the realization of interests under the Convention on the Law of the Sea—the needs of countries in the development and management of marine resources—comprehensively outlined the connotation of comprehensive marine management. In 1992 the publication of the Agenda 21 marked the formal establishment of the idea of comprehensive marine management.

1.2.2.2 Meaning of comprehensive marine management

Initially, the meaning of comprehensive marine management was relatively simple, referring to the overall consideration of human economic development activities, marine resources, and other elements in a specific region (Song, 2006; Stern, Common, & Barbier, 1996). The meaning of comprehensive marine management is further clarified in the Agenda 21; it refers to the sustainable utilization of marine resources and the improvement of the marine environment through legislation, policymaking, and organizing, and by coordinating with relevant departments. Since the implementation of the Agenda 21, many scholars have further supplemented the meaning of comprehensive marine management, which can be summarized as follows (Eikeset, Mazzarella, & Davsdttir et al., 2018; Lange & Jiddawi, 2009; Masalu, 2000; Sherman, Patricia, Alvarez, & Peterspm, 2017):

1. The comprehensive marine management uses management functions, such as planning, organization, leadership, and control to manage marine resources, environment, and rights and interests.
2. The goal of comprehensive marine management is to develop the overall function of oceans and create basic conditions for the sustainable utilization of marine resources for the overall interests.
3. The comprehensive marine management focuses on integrity, comprehensiveness, and scientificity, which falls under the systematic senior management strategy and seldom involves specific industries and economic activities.

1.2.3 Theory of sustainable utilization of marine resources

1.2.3.1 Theory of marine safety

In 1980 to adapt to the gradual improvement of productivity and living standards, safety science theory as a new research field emerged and continued to expand (Hodson & Marvin, 2009; Lawton, 2007; Mcdonald, 2018; Zhao, et al., 2006). Marine resource safety management is a systematic and multidisciplinary subject. The marine safety theory provides a solid theoretical basis for achieving the sustainable utilization of marine resources (Lange & Jiddawi, 2009; Perivoliotis, Krokos, Nittis, & Korres, 2011).

In a narrow sense, the theory of marine safety mainly refers to the safety of demand and supply of marine resources, focusing on the management of marine resources reserves. At the level of marine resource awareness, there are many concepts of marine resources, such as resource shortage, resource development, and the conservation and protection of marine resources. At the level of the marine resource system, it includes legislation, policy formulation, and systems construction for marine resource management. At the material level of marine resources, it mainly includes the development and utilization, production and consumption, and transportation and storage of marine resources.

The generalized marine safety theory can be divided into the following three layers: first, the supply capacity of marine resources must meet the production and living needs of a certain region, that is, to ensure the "quantity" of marine resources; second, the development and utilization of marine resources should ensure the sustainable development of the marine environment, that is, to ensure the "quality" of marine resources; third, the market price of marine resources should be lower than the affordable range of residents' consumption capacity, that is, to ensure the "price" of marine resources.

1.2.3.2 Theory of economic externality

The theory of economic externality was originally put forward by Marshall. The essence of this theory is to reflect that, in market economy, the parties do not bear all the consequences of their own economic activities; in other words, to the theory reflects the spillover effect of economic activities (Marshall, 1890). The externality theory of resource utilization reflects the fundamental reason for the low allocation efficiency of resources under the market economy conditions. It is independent of decision-making control and market mechanism and is unavoidable and difficult to eliminate.

Under the premise of the externality of the market economy, the market often fails, that is, the market price cannot regulate the efficient allocation of resources, which keeps the allocation of resources from reaching the Pareto optimum (Brock & Xepapadeas, 2010; Capello & Faggian, 2002). On the one hand, when the parties engage in relevant economic activities, there is always a certain spillover of economic benefits, while the beneficiaries of indirect economic benefits do not need to pay fees, which makes the economic effects low. On the other hand, when the economic activities of the parties show negative externalities, that is, the economic activities cause economic losses to others, the parties are not required to compensate for the losses. Eventually, this leads to inefficient economic efficiency.

Under the above circumstances, it is far from enough to rely solely on the market mechanism to adjust the allocation of resources. Therefore forces outside the market are needed. The common mechanisms include public opinion and administrative intervention (Helmsing, 2001; Zuo, 1994), among which administrative intervention can be divided into the following five aspects: legal control, the collection of sewage discharge fees, sewage tax, sewage discharge license, and clear property rights.

1.2.3.3 Marine ecological carrying capacity

The marine ecosystem includes the marine abiotic environment and the marine biological environment. They mutually affect each other through the circulation of material and energy, thus developing a system with a self-regulation ability (Di, Han, Liu, & Chang, 2007; Kang, Xu, & Jiao, 2013). Due to the influence of the structure of marine the ecosystem, the carrying capacity of marine ecosystem encompasses the following multilayered relationship (Akpalu & Bitew, 2011; Miao, Wang, Zhang, & Wang, 2006; Ma, You, Ma, Xie, & Li, 2012; Wang, Wang, & Song, 2017):

1. The basic layer mainly includes seawater resources, light, salinity, and temperature, which aim to provide a basic living space for marine organisms.
2. The secondary layer mainly includes marine phytoplankton and organisms, which aim to provide oxygen and purified water and to enrich the diversity of marine organisms.
3. The tertiary layer mainly contains high-level organisms in the ocean, which aim to provide food, enrich the diversity of marine organisms, and stabilize the food chain.

1.3 Summary of the current situations of sustainable marine resource utilization

1.3.1 Japan

Japan is a powerful maritime country. In order to obtain marine benefits, Japan's marine strategy depends largely on its naval power. After the Second World War, the Japanese government adjusted its national economic development strategy, and nonmilitary factors, such as marine resources and environmental protection, became the focus of Japan's marine strategy. In 1996 the Japanese government approved the establishment of the exclusive economic zone, declaring that the total area of its exclusive economic zone was 4.47 million km^2 (including disputed areas with neighboring countries), ranking sixth globally. In the 21st century, witnessing an increase in maritime issues, including the expansion of maritime jurisdiction and disputes over maritime rights, the Japanese government began to adjust its marine development strategy and marine management system by formulating new marine policies and establishing and improving relevant systems.

Japan's marine management system is decentralized. Since there is no special department responsible for marine affairs, the work related to oceans is scattered among the following government departments: the Ministry of Land, Infrastructure, Transport and Tourism; Ministry of Agriculture, Forestry and Fisheries; Land Bureau; and Natural Resources and Energy Agency. In 2001 Japan integrated the relevant departments and the marine work was undertaken by the Ministry of Land, Infrastructure, Transport and Tourism and the Ministry Agriculture, Forestry and Fisheries, and the Ministry of Environment. Although institutional arrangements have shrunk, decentralized management exists (Wakita & Yagi, 2013). To coordinate the policies among various marine departments and formulate marine development plans, the Japanese government gradually established various coordination departments according to its own needs. For example, the Commission for the Promotion of Marine Science and Technology was established in 1969 to coordinate and plan marine development among different ministries. Although Japan has a sophisticated marine coordination department, due to the lack of a unified management department, there are many loopholes in the comprehensive marine management, such as department overlap and interdepartmental conflicts. This implies the possibility of inefficient coordination among relevant departments regarding marine issues. To solve these problems, the Japanese government established the Headquarters of Marine Policy in 2007, with the Prime Minister as the Department Minister, the Chief Cabinet Secretary, and the Minister of Domestic Transport as Vice-Ministers.

In addition, Japan has a strong understanding of marine management, especially pertaining to the prevention and control of marine pollution. Since the 1950s, Japan has promulgated a series of laws to control marine environmental pollution, including the Coastal Law (1953) and the Seto Law on Special Measures for Inland Marine Environmental Protection (1978). Since the 1990s, under the restraint of the Convention on the

Law of the Sea, the Japanese government gradually established a marine legal system, including the Exclusive Economic Zone and Continental Shelf Law (1996), the Exclusive Economic Zone Fishing Exclusive Law (1996), and the Basic Action of Marine Policy (2007). The Basic Regulations on Marine Policy (2007) clearly defines that Japan's basic marine policy must protect and guarantee marine safety, enrich marine scientific knowledge, develop marine industry, integrate marine and coastal management, and conduct international cooperation.

1.3.2 South Korea

South Korea's earliest maritime management departments were the Shipping Authority (Ministry of Communications) and the Fisheries Authority (Ministry of Commerce and Industry), which were both established in 1948; these departments exemplified the decentralization of maritime management. Subsequently, in 1955, South Korea established a maritime office for ports, fisheries, shipbuilding, and coast guards, which was regarded as the early centralization of maritime administration. However, in 1961, the Korean government divided the maritime office into 13 marine-related departments, including the harbor office, the Fisheries Bureau, the Ministry of Science and Technology, the Ministry of Agriculture and Forestry, and the Ministry of Industry. This adjustment indicated the decentralization of South Korea's marine management, which lasted for 30 years. After this period, South Korea reformed the marine management and established the Ministry of Ocean Affairs and Fisheries (MOMAF) in 1996 (Hong & Chang, 1997). After the promulgation of the Rio Declaration and the Agenda 21 in 1992, the Korean government recognized the importance of implementing comprehensive marine management and sustainable marine development and utilization and became one of the earliest countries to implement centralized marine management. However, it is noteworthy that the Korean centralization lasted only 12 years. In 2008 President Myung-bak Lee withdrew from the Ministry of Oceans and decentralized his functions to the Ministry of Land, Transport and Maritime Affairs, and the Ministry of Agriculture, Forestry, Fisheries and Food, thereby weakening the central government's marine leadership and control functions (Liu, Ballinger, Jaleel, Wu, & Lin, 2012). In 2013 the Marine Authority was reestablished. After several years of practice, the centralization of marine management in Korea led to remarkable achievements.

Before the 1990s, Korea had enacted many laws related to marine management, such as the Sewage Clearance Act (1961), the Public Water Management Act (1961), the National Land Comprehensive Planning and Construction Act (1963), the Marine Pollution Prevention, the Control Act (1977), and the Basic Law on Marine Development (1987). Since the 1990s, especially since the United Nations Conference on Environment and Development, Korea adjusted its marine strategy accordingly and made great progress in formulating marine laws and policies. Several laws related to marine environmental management have been promulgated, including the Coastal Management Law (1999), the Wetland Protection Law (1999), the Basic Law on Marine and Fisheries Development (2002), and the Marine Environmental Management Law (2007). It is worth noting that legislation is an important part of a country's legal system; thus the centralized and decentralized modes of marine management can be distinguished according to the essence of legislation. When the substantive content of legislation includes the establishment of local systems or the formulation of different policies according to regional conditions, it means that the mode of marine management is not centralized.

1.3.3 Russia

Russia is an oceanic power that constantly opens up its sea area. After Putin took office as President, its marine economic level has been continuously improving, and its marine issues are gaining attention. On July 27, 2001, the Russian President approved the "Oceanography of the Russian Federation for the Period 2020" (2001), which is the first programmatic document formulated by Russia for the full implementation of its national marine policy in many years. It not only provides the legal basis for Russia's marine policy, but also fundamentally unifies the marine management rights of different divisions. Specifically, the Russian marine management system mainly includes the Federal President and relevant government departments, such as the Russian Federation Conference, the Ocean Commission, the Ministry of Agriculture, the Ministry of Communications, the Ministry of Natural Resources, and local administrative departments (Wan & Chen, 2014). Among them, the Federal President is responsible for formulating short- and long-term marine strategies, safeguarding marine interests, and guiding the implementation of policies. The Russian Federation Conference is responsible for legislative work to ensure the implementation of relevant policies. Relevant marine management departments are responsible for the implementation of the policies. The functions of marine departments mainly include the following: the Ministry of Natural Resources undertakes the exploration, development, and protection of marine resources; the Ministry of Communications undertakes maritime transportation; and the Ministry of Agriculture undertakes fisheries management. In addition, the Ocean Commission is mainly responsible for coordinating the work of relevant departments to ensure the smooth progress of marine management.

Russia's national marine strategy is mainly based on the Russian Federation's special marine outline before 2020 (hereinafter referred to as the Outline). The outline was formally approved by the Government of the Russian Federation in its resolution 919 on August 10, 1998, and was frequently revised until May 2006. The basic purpose of the outline is to solve a series of problems, including development, the effective use of the world's marine resources and space to ensure national security, and economic interests. Meanwhile, the development and utilization of marine resources is an important source for Russia, and hence it must preserve and expand its reserves of raw materials. To maintain its economic and production independence, Russia attaches great importance to the protection of marine rights (Lin & Mi, 2009). In accordance with the United Nations Convention on the Law of the Sea promulgated in 1994, Russia promulgated three important laws on oceans in 1995 and 1998. In 2001 Russia defined its maritime rights, the rights of marine minerals and biological resources from a legal point of view, which expressed Russia's position on the development and utilization of marine resources.

1.3.4 The United States of America

The lighthouse service established by the United States of America in 1716 was the prologue of marine management in the country (Shen, 2016). Since then, the marine management in the United States of America is mainly based on the administrative division management mode. Until the 1960s, facing the energy crisis and various economic problems, the United States of America established the National Marine Resources and Engineering Commission (also known as the Strardons Commission), which mainly manages all major marine activities. It reflects the increased intervention of the Federal Government in state marine management. In the 1970s, the country further established the National

Oceanic and Atmospheric Administration (NOAA), an independent government agency responsible for national ocean and resources management, marine protection, formulation of national ocean-related policies, and participation in the national ocean affairs and cooperation. In the 1990s, with the further development and utilization of marine resources, to improve the existing marine management system, the United States Congress established the National Ocean Policy Committee in August 2000 under the Ocean Act, which is responsible for the comprehensive marine agenda of the 21st century. In 2004 the United States of America established the Environmental Quality Committee in the Marine Policy Committee (Cabinet). Since then, to improve the marine management system and strengthen the coordination of various departments within the marine management system, the Ocean Commission of the United States of American has established various branches, such as the Inter-ministerial Marine Science and Resource Management Integration Committee, to coordinate the sea-related affairs. With the change in the economic situation at home and abroad, the Obama administration established a new National Ocean Council (NOC) in 2010 to strengthen the marine management of the United States of America based on a series of integration of existing marine management agencies.

As the first country in the world to formulate ocean planning, the United States of America formulated the Oceanographic Decade Plan (1960–70) as early as 1959, and, in the same year, formulated the world's first military marine planning, the Naval Oceanographic Decade Plan. Since then, the United States of America has been launching a series of marine development plans, including the Long-term Oceanographic Plan of the United States (1963–72; 1963), National Marine Science and Technology Development Plan (1986; Yin, Wei, & Meng, 2009), Coastal Ocean Planning (1989), Marine Science and Technology Development Report of the 1990s (1990), and the United States' Marine Agenda 21 (1998). In 1999 the United States of America formulated a national marine strategy and a marine economic plan. Subsequently, the Oceans Act of 2000 and the strategic plan for marine development in the 21st century were formulated.

The changes in the marine ecological environment and economic development situations have driven every country to emphasize the importance of sustainable development. Considering the sustainability and regeneration of the exploitation and utilization of marine resources, the United States of America promulgated the 21st Century Ocean Blueprint in 2004, which provides the most comprehensive assessment of all aspects of the oceans to date. Thereafter, to effectively implement the 21st Century Ocean Blueprint, the country promulgated the US Ocean Action Plan in the same year, which defined the specific strategic path for the implementation of the plan. With the gradual development of marine industries, the US government successively promulgated the planning for the ocean science cause of the United States for the next 10 years. It included the Priority Plan and Implementation Strategy for Marine Research (2007) and the memorandum of understanding for formulating the US marine policy and implementation strategy (2009), which further improved the marine planning system of the United States of America.

1.3.5 Britain

As far as the construction of marine resource management institutions is concerned, there is still no unified government department for managing ocean affairs in the United Kingdom. The different affairs of marine management are scattered among various administrative departments, including not only government departments but also some semiofficial organizations licensed by the government. An increase in the number of marine affairs led to a gradual improvement in the management system of marine resources in Britain. Despite this gradual improvement, Britain witnessed a decentralized management system of marine resources. Owing to the clear division of labor and the high efficiency of coordination management, this marine management system is more effective (Wang, 2012). In addition, to deal with the contradictions among different departments more efficiently, Britain also established the Marine Science and Technology and Marine Property Commission.

As an island country, dependent on the development of the marine economy, the core aim of British marine management is to strengthen the legislation. The characteristic of this management mode is to make corresponding legal documents for different ocean affairs. In addition, each department is independent but mutually supportive. In other words, it does not rely solely on a comprehensive regulation for management. Over the years, the development of marine resources in Britain has been mainly focused on obtaining long-term economic benefits. This shows that Britain attaches great importance to the coordination between economic development and marine environmental protection in the process of marine legislation. It requires that the sustainability of marine resources should be considered while developing and utilizing marine resources. Therefore at the beginning of the 21st century, Britain also established the Marine Management Bureau; this Bureau regularly assesses the conditions of marine self-environment, exploits and utilizes marine resources, and provides strong support and guarantee for sustainable marine development.

Before 2000, the marine resource management policies formulated by the United Kingdom were decentralized and were based on a single industry or region. Specifically, with regards to marine resource development, Britain promulgated the Offshore Oil Development Act (1975), the Oil Act (1988), and the Fisheries Act (1981). In the field of marine environmental protection, Britain promulgated the Coastal Protection Act (1949) and the Oil Pollution Prevention Act (1971). Concerning the ownership of maritime areas and real estate planning, Britain promulgated the Royal Real Estate Act (1961), the Continental Shelf Act (1964), the Urban and Rural Planning Act (1971), and the Territorial Sea Act (1987). The above regulations constituted the basic legal system for the development and management of marine resources in Britain before the 20th century.

In the 1990s, the formulation of comprehensive marine policies attracted great attention from the British government, marine conservation organizations, the scientific and technological communities, and the general public. In 2001 Britain focused on the formulation of comprehensive and high-level marine management policies. Subsequently, in 2002, the Department of Ocean, Food and Rural Affairs of the United Kingdom proposed that the goal of the United Kingdom in the field of oceans is to facilitate "a clean, healthy, safe, productive, and bio-diversified ocean," and issued a report "Protecting Our Oceans" accordingly. In 2003 the British government issued the report named "Changing Oceans," which recommended the formulation of comprehensive marine policies and the utilization of new management methods to manage all types of marine activities. In 2008 Britain issued the "2025 Marine Science and Technology Plan," which proposed that priority should be given to supporting biodiversity and sustainable use of marine resources in 10 different fields. In 2009 Britain formally ratified the Law of the Sea. In 2010 Britain released the Marine Energy Action Plan, which proposed to support the maritime development in technology, finance, policy, and other maritime aspects.

1.3.6 Australia

As one of the countries in the world with the largest sea area under jurisdiction (Fu, 2013), Australia's marine resource management is closely related to the country's socioeconomic development, which makes Australia's marine functional institutions and work-coordination mechanism very comprehensive. Australia has become one of the first countries in the world to adopt regional Ocean planning and implement marine policies. As the main administrative body of Australia, the Commonwealth of Australia and the governments of various states made many useful attempts in promoting marine rights, environment, resources, and international marine affairs. Ultimately, these efforts led to the establishment of the marine management mechanism of the division of labor and cooperation between the federal government and the state governments. The Federal government is responsible for foreign affairs, defense, immigration, and customs affairs, whereas the state government is responsible for other marine affairs in their water areas (Wang, 2012).

Australia has also established the national oceanic ministerial committee to strengthen the unified leadership of marine management and coordinate the management of oceans. However, with the change of government and the adjustment of the concept of governance, the national oceanic ministerial committee disintegrated in 2005. Since then, the coordination of marine affairs in Australia has been undertaken by the committee of ministers of natural resources management. It was established in 2001 and is responsible for monitoring, assessing, and reporting Australian marine affairs, especially natural resources management. Australia's maritime powers are relatively dispersed in various ministries and commissions. Similar to Britain, Australia's maritime management system is also decentralized.

In addition, to accelerate the exploration of continental shelf resources to meet the needs of development, Australia implemented the Continental Margin Depth Mapping Plan in 1989, which maps the continental margin according to the depth and topography of the continental margin. In 1990 the formulation of Australia's Marine Industry Development Strategy (1990—94) not only supported the management of marine resources but also safeguarded the sovereignty of territorial waters and marine resources. In response to changes in the economic situation, social structure, and ecological environment, Australia formulated the Marine Industry Development Strategy (1997) and Marine Science and Technology Development Plan (1998), respectively.

1.3.7 France

In 1960 the French President Charles de Gaulle proposed that France should march to the sea. In 1967 the French National Centre for Marine Development was established as a public research institution with both industrial and commercial characteristics. Its main task is to connect state-owned and private enterprises, develop marine-related science and technology, and explore the development and utilization of marine resources. In the early 1970s, to make the maritime consciousness more popular, France formulated the strategic goal to emerge as a great sea power, namely, to increase oceanographic survey and make full use of marine resources.

In the 1980s, France accelerated its marine strategy process, keeping in mind that "the oceans bring wealth and risks, and it is imperative to strengthen marine management and control." Therefore France attaches great importance to the regulatory role of the government and concentrates on the unified regulatory mechanism to implement its marine strategy. Hence, France has established the Institute for Marine Development. In May 1981, France set up the Ministry of Oceans and later changed it to the State Secretariat of Oceans, which is an important department that ensures unified supervision and coordination of ocean work. It is mainly responsible for the formulation and implementation of French marine policies, which pioneered the establishment of such institutions in Western countries. The establishment of the State Secretariat for the Oceans significantly promoted the development of French marine undertakings and the realization of France's goal of centralized and unified regulation of the oceans. In order to implement the Agenda 21 and the European Union Green Paper on Marine Policy and the Blue Paper on Integrated Marine Policy, France implemented a series of comprehensive management measures to realize the development and management of marine resources. In the early 21st century, a high-level committee of experts on oceans was established, which was mainly responsible for the formulation of marine development policies in the next decade. These practices have clear orientation and comprehensive supporting paths, which is of great practical significance for China as it plans to implement strategies for boosting its maritime power.

1.3.8 Canada

Canada is a large maritime country where oceans are very important for its survival and development. According to its Constitution, the Federal

Government is mainly responsible for the management of marine affairs in Canada, and the boundary between the Federal Government management and provincial government management is the low tide line. Provinces have jurisdiction over resources within the low tide line, while resources beyond the low tide line are managed by their respective relevant Federal Government departments according to industry divisions. The Department of Marine Fisheries, established in 1979 as the competent department of Canada's National Marine Fisheries affairs, is responsible for coordinating federal policies and plans related to the oceans. It consists of the Canadian Coast Guard, the Fisheries and Agriculture Administration, the Marine and Eco-Environment Administration, and relevant marine scientific research centers. In 1997 Canada established the Law of the Sea to implement changes in the international maritime situation after its entry into the United Nations Convention on the Law of the Sea in 1982, and it became the first country in the world to implement the comprehensive marine legislation. The Law of the Sea clearly divides the administrative boundaries and powers of the Federal and provincial governments and explains the administrative responsibilities of various departments of the Federal Government.

In addition to the mentioned Law of the Sea, Canada's marine resource management legislation includes the following: the Coastal Fisheries Protection Act (1869) deals with the monitoring and control of marine living resources; and the Fisheries Law (1868) provides for the protection of marine ecological environment and marine living resources; the scope covered by the Fisheries Development Law mainly covers the utilization, protection, and development of marine resources. Since Canada's entry into the Law of the Sea, to optimize the marine resource management system, Canada has been implementing relevant policies. For example, in order to promote marine environmental protection, the Canadian Federal Government formulated the Green Plan at the end of the 20th century. Meantime, the Atlantic Coast Action Plan and Habitat Action Plan were introduced.

1.3.9 Vietnam

Unlike most countries, Vietnam implements an industry-based marine management approach, which is managed by different departments, institutions, and local authorities, according to division of labor. For example, the Ministry of Fisheries is responsible for the management of fisheries and biological resources, the Ministry of Energy is responsible for the development and management of offshore oil and gas resources, the Ministry of Communications is responsible for the management of ports and shipping, and the Ministry of Science and Technology and the Ministry of Environment are responsible for the management of the marine environment. Vietnam mainly formulates maritime laws and regulations on the basis of the Declaration on the Determination of the Baseline Width of the Territorial Sea (1982) and the Declaration on Territorial Sea, Adjacent Areas, Exclusive Economic Zone, and Continental Shelf (1997). In addition, its legal system establishes the legal status of the maritime jurisdiction, mainly including the Petroleum Law (1970) and the Regulations on the Protection of Aquatic Resources (1987).

Vietnam attaches great importance to policy formulation for the management of marine resources. To support the development of deep-sea fisheries, Vietnam has formulated an ocean-going sea fishing plan and promulgated relevant preferential tariff regulations. In 1998 to promote the development and utilization of oil and gas, Vietnam formulated preferential conditions for natural gas and oil investment activities. In 1999 to promote the development of mariculture, Vietnam launched the Aquaculture Plan (1999—2010). In 2000 Vietnam drafted the national marine protected areas plan to protect marine resources. In 2003 Vietnam formulated the fisheries facilities plan (2003—10) and invested USD 130 million to build and improve the core marine aquatic facilities in southern Vietnam. In 2007 Vietnam adopted the Vietnamese Ocean Strategy Resolution 2020, which combines marine management, economic defense, and politics. To provide an important tool for the implementation of maritime strategy, Vietnam promulgated the Vietnamese Law of the Sea in 2012, which makes its marine management system more perfect and efficient.

1.3.10 China

Oceans have long been considered as one of the most important natural resources of mankind (Costanza, 2000). As one of the earliest countries to exploit and utilize oceans, China has a history of developing and utilizing marine resources for thousands of years, and has advanced marine salt industry, fishery, marine transportation industry, and other industries. However, the lack of coastal defense, the outbreak of the Opium War between China and Britain, and a series of subsequent marine aggression led China to a distressful phase and also taught the country valuable lessons. Since the founding of New China, the development and management of marine resources has gradually been on the right track. The strategy of building China into a powerful marine country is being implemented phase wise.

From 1949 to 1960, China's marine undertakings began to develop in an all-round way. The management of marine resources was constantly strengthened, and the management level was constantly improved. According to the actual situation of marine resources in China at that time, marine resource management continued to follow the management model of land industry. Marine resource and land resource industries were managed together, and management departments were set up for marine fisheries, marine transportation, and marine salt industries, respectively. China's marine management model was first established, which laid an important foundation for the early development of the new China's maritime economy.

Subsequently, from 1964 to 1979, China's marine industry developed rapidly. The outputs of the shipbuilding, fishing, or aquaculture industries have been registering a significant increase, and the area of sea salt fields has been expanding. In this case, China's marine resource management system needed further modification and improvement. Thus the establishment of the State Oceanic Administration was approved at the 124th Standing Committee of the Second NPC in 1964, which was under the jurisdiction of the navy department at that time. The State Oceanic Administration set up branches in Qingdao, Ningbo, and Guangzhou provinces with corresponding marine information and data centers and marine instrument research institutes. The establishment of the State Oceanic Administration marked the beginning of a more stable management system of marine resources in China.

After the reform and opening up, the administration of marine resources in China witnessed gradual unification. In 1982 the State Oceanic Administration was established as a subordinate organ under the State Council. It is responsible for organizing and coordinating ocean work throughout the country. It mainly undertakes the responsibilities of scientific research, management, investigation, and public service in the national ocean field. In a subsequent development, a four-level management system comprising the State Oceanic Administration, the sub-bureau of Marine Areas, the Marine Administration Area, and the Marine Supervision Station was formed. This system continuously contributed toward the development of China's marine undertakings and the construction of maritime frontier defense. Later, since coastal areas gradually showed superior geographical and resource advantages after the reform and opening up, the management of local marine resources gained the attention of marine resource management. In 1980 in order to fully grasp the coastal zone and marine resources of the whole country, the State Oceanic Administration and various departments set up special research groups and "coastal zone investigation office" in coastal areas, which further strengthened China's marine management. With the further development of China's marine management system, coastal provinces, and municipalities set up marine management agencies.

In 2002 China began to implement the Law of the People's Republic of China on the Utilization and Management of Sea Areas, which further clarified the power relationship between the State Oceanic Administration and local marine administrative agencies in coastal areas. China's marine management showed a distinct hierarchical pattern. In 2003 China's first medium- and long-term plan for the development of the national marine economy began to be implemented. Meanwhile, the State Council issued the Outline, which defined the basic framework and direction for the implementation of the marine development strategy. Subsequently, the State Council approved the establishment of marine functional zones in Zhejiang and Fujian in 2004 and issued the Outline of the National Marine Development Plan in 2008, aiming at regulating and strategizing for the marine economy, and strengthening marine management. In 2011 the State Council approved and implemented the Development Plan of Blue Economic Zone in Shandong Peninsula. The implementation of this plan marked the progress of China's marine management system, which gradually turned toward localization.

Although relevant provincial departments are responsible for comprehensive marine management, due to the lack of higher authority, management finds it difficult to achieve integration. Similarly, the lack of cross-sectoral coordination system in planning and management often results in the untimely handling of ocean affairs (Cao & Wong, 2007; Shi, Hutchinson, Yu, & Xu, 2001). Although these governments tried to integrate, sectoral self-interest and regional egoism overthrew these coordination systems (Chou, 2005). Therefore to improve the marine management, the Chinese government restructured the State Oceanic Administration in 2013. It integrated the former State Oceanic Administration, Marine Monitoring, Coastal Guard of the Ministry of Public Security, Fisheries Supervision of the Ministry of Agriculture and

General Administration of Customs. After the reorganization, the marine authority was transferred to the restructured State Oceanic Administration. The State Council stipulates that the restructured State Oceanic Administration shall be administered by the Ministry of Land and Resources. Its main tasks are to draft marine development plans, implement marine rights, manage marine utilization, and protect the marine environment. The restructured State Oceanic Administration implements maritime rights in the name of the China Coast Guard and receives operational guidance from the Ministry of Public Security. Since the reorganization, the State Oceanic Administration has strengthened the supervision and enforcement of marine management. Meanwhile, China established the Ocean Commission, which is a special institution for studying national marine development strategies and coordinating major marine affairs. The specific work of the Oceanic Commission is undertaken by the State Oceanic Administration. Until now, China's marine management has gradually changed from decentralization to centralization.

The promulgation of the Laws of the Sea in China started relatively late. In the 1970s, the government formulated provisional regulations for the prevention of pollution in coastal waters (1974). Since the 1980s, the legal system of ocean management has been gradually formed with a series of sea laws and policies. There are more than 10 laws related to marine management, including the Environmental Protection Law (1989, revised in 2014), the Soil and Water Conservation Law (1991, 2010), the Marine Environmental Protection Law (1999, revised in 2014), and the Sea Area Utilization Management Law (2001). The other marine policies issued included China's Agenda 21 for Oceans (1996), the Focus and Distribution of Marine Industry Development (2011−20; 2010), and the Twelfth Outline of the Five-Year Plan for National Economic and Social Development (2011−15).

1.4 Research contents and programs in this book

1.4.1 Research thoughts

This book has the following sequence: extracting key issues, elaborating basic theories, analyzing realistic conditions, evaluating utilization benefits, assessing utilization effects, explaining management effects and designing policy system, aiming at making breakthroughs on theoretical research, acquiring feasible countermeasures that can effectively support the sustainable utilization, and managing evaluation of marine resources in China. In other words, first, the book extracted the core problems that must be solved urgently to achieve sustainable development, utilization, and management of marine resources. Second, the book evaluated the carrying capacity of China's marine ecosystem and analyzed the basic conditions for sustainable utilization of marine resources in China's coastal areas. Third, the book examined the economic, environmental, and comprehensive benefits of marine resource utilization in China and clarified the realistic basis for sustainable utilization of marine resources in China's coastal areas. Fourth, the book analyzed the coordinated coupling relationship among the utilization of marine resources in China, economy, and environment and grasped the social impacts of sustainable utilization of marine resources in China's coastal areas. Finally, the book evaluated the management level of marine resources in China and explored the differences in the quality of sustainable utilization and management of marine resources in the coastal areas of China.

1.4.2 Research contents

Chapter 1, An overview of sustainable marine resource utilization, summarizes the sustainable utilization of marine resource. This chapter aims to lay a foundation for the whole research. First, the research background and theoretical basis of this book are presented. Subsequently, practical management experiences in maritime states globally are summarized, and the basic problems to be researched in this book are abstracted. Further, the overall framework of this chapter is established through the explanations on research ideas, logical framework, and the chapter arrangement. Finally, research methods and the technical route adopted in this chapter are discussed.

Chapter 2, Assessment of China's marine ecological carrying capacity, assesses China's marine ecological carrying capacity. First, the index system and the model for evaluating the marine ecological carrying capacity are constructed. Subsequently, the spatial differentiation in China's marine economic sustainable development capacity is clarified using relevant knowledge, such as geo-economics, and the relationship between ecological carrying capacity and industrial structure based on correlation analysis is further discussed.

Chapter 3, Analysis of influencing factors and efficiency of marine resource utilization in China, analyzes the efficiency and influencing factors of China's marine resource utilization. Considering current situations of marine resource utilization in China and the availability of data, corresponding indexes for the accounting of marine resource utilization efficiency are obtained. Subsequently, based on the analysis framework of super-efficiency SBM-Undesirable and metafrontier models, the efficiency of marine resource utilization is measured. Meanwhile, the dynamic evolution law of marine resource utilization efficiency is estimated using nuclear density estimation and Markov chain estimation methods, and the factors influencing the marine resource utilization efficiency are evaluated using the symbolic regression method.

Chapter 4, Analysis of the marine carbon sink capacity in China, analyzes China's marine carbon sink capacity. First, this chapter clarifies the mechanism of ocean carbon cycle and carbon dioxide absorption in the ocean and proposes an estimation method for the whole large-scale ocean carbon sinks. Subsequently, based on the relevant data of marine economy of coastal areas in China, an empirical analysis of marine carbon sink capacity is conducted.

Chapter 5, Comprehensive benefit evaluation of marine resource utilization in China, conducts a comprehensive benefit evaluation of marine resource utilization in China. This chapter comprehensively considers the impacts of population, social economy, resources, industrial structure and ecological services. It also establishes a comprehensive benefit index system covering economic, social, and ecological benefits of marine resource development and utilization. Subsequently, the system dynamics model and mean square error weight method are used to quantitatively evaluate the comprehensive benefits of marine resource utilization in China's coastal areas from 2006 to 2015.

Chapter 6, Decoupling marine resources and economic development in China, analyzes the decoupling between China's marine resource and economic development. Using the theory and method of ecological footprint, the composition of marine resource footprint of coastal areas in China is calculated, and the coordinated relationship between marine resource utilization and economic growth is evaluated by considering the decoupling evaluation model of coordinated development.

Chapter 7, Analysis of coupling among marine resources, environment, and economy in China, analyzes the coupling relationship among marine resources, environment, and economy in China. In this chapter, based on the establishment of a comprehensive evaluation index system for coordinated development among marine resources, environment, and economy, the coupling degree model is used for spatial analysis of the coupling degree among marine resource endowment, ecological environment, and economic development in China's coastal areas.

Chapter 8, Evaluation of marine resource management levels in China, is about China's marine resource management level. The evaluation system of China's marine resource management level is constructed based on the following three aspects: scientific and technological management, environmental management, and administrative management. In addition, a comprehensive evaluation method combining the entropy weight method and expert evaluation method is constructed. Subsequently, the marine management level of coastal areas in China is evaluated quantitatively.

1.4.3 Research methods

To comprehensively address marine problems, this chapter proposes the use of appropriate methods in resource economics, regional economics, marine ecology, mathematical economics, and other disciplines during the actual research process. Specifically, the following methods are involved:

1. Literature retrieval method: The research basis of this book is laid by searching the research results related to the sustainable development of marine economy, summarizing research perspectives, methods, and conclusions in existing literatures, as well as the problems and countermeasures of China's marine economy development from different perspectives.

2. Logical analysis method: In the theoretical research process of this book, we used many kinds of logical analysis methods, such as classification, induction, deduction, and derivation, to put forward our own views. Based on previous studies, the concept and category of marine resource management are defined and classified, and the

theories of safe utilization of marine resources, economic externality, and ecological carrying capacity involved in sustainable utilization of marine resources are summarized. Subsequently, the problems and causes of sustainable utilization of marine resources are theoretically deduced.

3. Empirical analysis: Based on a theoretical analysis, this study uses a variety of quantitative analysis methods to analyze the efficiency evaluation of China's marine resource utilization, the coordinated development of marine resource—economy—environment, and the evaluation and effect of marine resource management. Specifically, in the following chapters, we first construct the evaluation index of China's marine ecological carrying capacity and analyze the relationship between China's marine carrying capacity and industrial structure. Subsequently, based on the total factor productivity framework, the utilization efficiency of marine resources is evaluated, and the potential factors affecting the utilization efficiency of marine resources are analyzed using the symbolic regression model. Further, based on the system dynamics model, the comprehensive benefits of China's marine resource utilization are analyzed; the multiindex evaluation system of carbon sink, coupling degree, and resource utilization decoupling is constructed; and the relationship between marine resource and external environment and economic and social activities is deeply analyzed. Finally, the management level of China's marine resources is evaluated based on the construction of the index system.

4. Normative analysis method: To make better use of marine resources, this demonstrates how to develop and utilize marine resources from the perspective of sustainable development, based on relevant theoretical research and empirical analysis. Specifically, it includes the formulation and implementation of marine resource utilization policies, the effective promotion of utilization efficiency of marine resources, and the strengthening and enhancement of the management functions of relevant departments.

1.4.4 Technical route

The technical route of this book is as shown in Fig. 1.1.

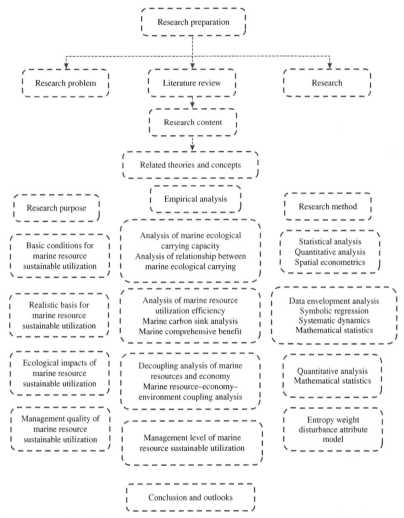

Figure 1.1 Technical route of the research.

1.5 Conclusion

This chapter outlines the concept of sustainable utilization of marine resources and related theories to provide theoretical reference for the follow-up study on the sustainable utilization of marine resources. Meanwhile, it gives a detailed overview of the practical experience of marine resource management in major marine countries. Subsequently, by considering the practical problems in sustainable utilization of marine resources and by taking China's marine resource management as an example, key issues and research contents to be solved in this book are put forward. As a comprehensive approach toward addressing these problems, this chapter proposes the use of appropriate methods in resource economics, regional economics, marine ecology, mathematical economics, and other disciplines.

References

Akpalu, W., & Bitew, W. T. (2011). Species diversity, fishing induced change in carrying capacity and sustainable fisheries management. *Ecological Economics.*, *70*(7), 1336−1343.

Brock, W., & Xepapadeas, A. (2010). Pattern formation, spatial externalities and regulation in coupled economic-ecological systems. *Journal of Environmental Economics and Management.*, *59*(2), 149−164.

Cao, W., & Wong, M. H. (2007). Current status of coastal zone issues and management in China: A review. *Environment International.*, *33*(7), 985−992.

Capello, R., & Faggian, A. (2002). An economic-ecological model of urban growth and urban externalities: Empirical evidence from Italy. *Ecological Economics*, *40*(2), 181−198.

Chou, C. H. (2005). *A study on China's marine affairs strategy and organization structure.* Taipei, China: Coastal Guard Administration of Executive Yuan.

Costanza, R. (2000). The ecological, economic, and social importance of the oceans. *Ecological Economics*, *31*(2), 199−213.

Di, Q., Han, Z., Liu, G., & Chang, H. (2007). Carrying capacity of marine region in Liaoning Province. *Chinese Geographical Science*, *17*(3), 229−235.

Eikeset, A. M., Mazzarella, A. B., Davsdttir, B., Klinger, D. H., Levin, S. A., Rovenskaya, E., & Stenseth, N. C. (2018). What is blue growth? The semantics of 'Sustainable Development' of marine environments. *Marine Policy*, *87*, 177−179.

Fu Y. (2013). Australia: A division of labour and a collaborative marine management mechanism. China Ocean News 4, 5−6.

Helmsing, A. H. J. (2001). Externalities, learning and governance: New perspectives on local economic development. *Development and Change*, *32*(2), 277−308.

Hodson, M., & Marvin, S. (2009). Urban ecological security: A new urban paradigm? *International Journal of Urban and Regional Research*, *33*(1), 193−215.

Hong, S. Y., & Chang, Y. T. (1997). Integrated coastal management and the advent of new ocean governance in Korea: Strategies for increasing the probability of success. *International Journal of Marine & Coastal Law*, *12*(2), 141−161.

Kang, Y. T., Xu, L. P., & Jiao, Y. (2013). Prediction models for marine traffic carrying capacity related resources and pollution. *Navigation of China*, *36*(1), 109−114.

Lange, G. M., & Jiddawi, N. (2009). Economic value of marine ecosystem services in Zanzibar: Implications for marine conservation and sustainable development. *Ocean and Coastal Management*, *52*(10), 521−532.

Lawton, J. H. (2007). Ecology, politics and policy. *Journal of Applied Ecology*, *44*(3), 465−474.

Lin, X., & Mi, G. X. (2009). Russian marine policy and strategy. *World Sci-Tech R&D*, *31* (01), 198−202.

Liu, W. H., Ballinger, R. C., Jaleel, A., Wu, C. C., & Lin, K. L. (2012). Comparative analysis of institutional and legal basis of marine and coastal management in the East Asian region. *Ocean Coast Management*, *62*, 43−53.

Ma, C. H., You, K., Ma, W. W., Xie, J., & Li, F. Q. (2012). A study on marine region carrying capacity and eco-compensation. *Journal of Ocean University of China*, *11*(2), 253−256.

Marshall, A. (1890). *Principles of economics*. Basingstoke, GB: Palgrave Macmillan.

Masalu, D. C. P. (2000). Coastal and marine resources use conflicts and sustainable development in Tanzania. *Ocean & Coastal Management*, *43*(6), 475−494.

Mcdonald, M. (2018). Climate change and security: Towards ecological security? *International Theory*, *10*(2), 153−180.

Miao, L. J., Wang, Y. G., Zhang, Y. H., & Wang, Q. M. (2006). Assessing index system for bearing capacity of marine ecological environment. *Marine Environment Science*, *25*(3), 75−77.

Perivoliotis, L., Krokos, G., Nittis, K., & Korres, G. (2011). The Aegean Sea marine security decision support system. *Ocean Science*, *7*(5), 671−683.

Shen, J. (2016). The experience and enlightenment of USA marine management and control. *China Maritime Safety*, *11*, 56−59.

Sherman, K., Patricia, M. S. N., Alvarez, T. P., & Peterspm, B. (2017). Sustainable development of Latin America and the Caribbean large marine ecosystems. *Environmental Development*, *22*, 1−8.

Shi, C., Hutchinson, S. M., Yu, L., & Xu, S. (2001). Towards a sustainable coast: An integrated coastal zone management framework for Shanghai, People's Republic of China. *Ocean & Coastal Management*, *44*(5−6), 411−427.

Song, C. S. (2006). Global challenges and strategies for control, conversion and utilization of CO_2 for sustainable development involving energy, catalysis, adsorption and chemical processing. *Catalysis Today*, *115*(1−4), 2−32.

Stern, D. I., Common, M. S., & Barbier, E. B. (1996). Economic growth and environmental degradation: The environmental Kuznets curve and sustainable development. *World Development*, *24*(7), 1151−1160.

Wakita, K., & Yagi, N. (2013). Evaluating integrated coastal management planning policy in Japan: Why the guideline 2000 has not been implemented. *Ocean & Coastal Management*, *84*, 97−106.

Wan, Q. S., & Chen, X. (2014). An analysis of Russian marine management system. *Russian Central Asia & East European Market*, *2014*(01), 51−61.

Wang, M. N. (2012). Institutional mechanism of maritime-related management of major marine countries and its enlightenment to China. *World Shipping*, *35*(03), 38−40.

Wang, S. H., Wang, Y. C., & Song, M. L. (2017). Construction and analogue simulation of TERE model for measuring marine bearing capacity in Qingdao. *Journal of Cleaner Production*, *167*, 1303−1031.

Yin, K. D., Wei, M. X., & Meng, Z. S. (2009). Development strategy and evolution of the world's major marine powers. *China Economist*, *4*, 8−10.

Zhao, Y. Z., Zou, X. Y., Cheng, H., Jia, H. K., Wu, Y. Q., Wang, G. Y., . . . Gao, S. Y. (2006). Assessing the ecological security of the Tibetan plateau: Methodology and a case study for Lhaze County. *Journal of Environmental Management*, *80*(2), 120−131.

Zuo, X. (1994). Socialist market economy and the family planning program in China: Some theoretical issues reconsidered. *Chinese Journal of Population Science*, *6*(3), 235−242.

Further reading

Qu, J. L. (1997). Developing the ocean cause and strengthening the study of ocean culture. *Journal of Qingdao Ocean University (Social Science Edition)*, *2*, 6−8.

CHAPTER TWO

Assessment of China's marine ecological carrying capacity

2.1 Introduction

Marine ecological carrying capacity refers to the maximum service function the system can provide on the premise of conserving the stability of the system. Marine ecological carrying capacity includes not only self-maintenance, self-regulation, and supply capacity of the ocean but also human activities such as production and living within the corresponding scope. China is a large country with a landmass of 9.6 million km^2 and an ocean area of only 3 million km^2. However, in recent years, due to the uncontrolled discharge of pollutants from industrial and agricultural activities, the indiscriminate exploitation of marine mineral resources, increased fishing intensity, and other reasons, the pressure of large-scale human activities surpasses the restoration capacity of China's marine ecosystem, causing serious damage to biological resources and the water environment. For example, due to frequent marine disasters, China suffered a direct economic loss of CNY6.398 billion in 2017. The destruction of marine ecological balance not only affects the health of marine systems but also sustained utilization of marine resources, which is a major challenge to China's development. In this context, it is of great significance to establish an effective evaluation index system to monitor the carrying capacity of the marine ecosystem.

However, the continued attention to the development and utilization of marine resources in China has supported the rapid development of the coastal economy, and the marine industry is growing steadily. According to the data released by the National Oceanographic Information Center, in the terms of economic scale, the average annual growth rate of China's total marine production is about 7.9% from 2010 to 2017. By 2017 China's total marine output value had reached CNY7.7611 trillion, making the marine economy more important in the entire national economic system. Currently, with the diversification of factors driving the marine economy and the strengthening of land−ocean economic relations,

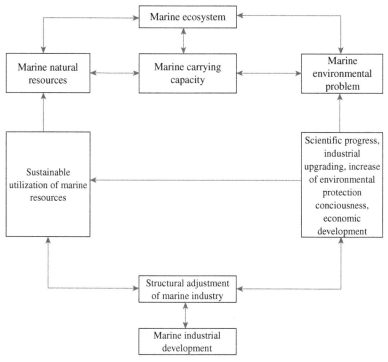

Figure 2.1 Relationship between marine carrying capacity and marine industrial structure.

marine economic growth in China is becoming increasingly significant. The key to realizing sustainable development of the marine economy is to form an efficient and balanced marine industrial structure; however, whether the marine industrial structure is reasonable depends on whether the functions are established according to the marine system's carrying capacity (Wang, 2013). It is equally important to scientifically measure the country's marine carrying capacity, suitability of ocean development, and to explore the relationship between ocean carrying capacity and industrial structure. The interaction between marine industrial structure and marine ecological carrying capacity can be expressed as in Fig. 2.1.

2.2 Literature review

The core content of research on marine ecological carrying capacity is to understand the benign cycle of marine ecological environment in coastal areas, ensure sustainable development of coastal areas, determine the optimal sustainable population growth rate, and social and economic development rate in coastal areas, according to the actual carrying capacity of the marine ecological environment and by deploying scientific methods to ascertain the balancing point between population growth, economic development, resource allocation, and marine ecological carrying capacity in coastal areas. Marine ecological carrying capacity reflects an interactive coupling relationship between human activities and the marine ecosystem as a carrier medium.

The concept of marine ecological carrying capacity is developed and separated from the traditional carrying capacity theory, which is often used to measure the relationship between human activities and the natural environment. The earliest record of carrying capacity started from human statistics at the end of the 18th century. With the continuous development and improvement of the carrying capacity theory, Park and Burgess (1920) first applied ecological carrying capacity to the field of human ecology. Later, the concepts of land resource carrying capacity, water resource carrying capacity, and environmental carrying capacity were proposed successively. In the 1970s Holling (1973) proposed the concept of ecological carrying capacity and established the conceptual/theoretical model. From the 1980s onward, research on ecological carrying capacity has developed further. Slesser (1990) proposed a new method for calculating the carrying capacity of resources and the environment, namely, a strategic model that improves the carrying capacity and considers the relationship between population and resources. He established a system dynamics model with long-term development goals to determine the optimal scheme of regional development by simulating different development strategies. At the beginning of the 21st century, Brown and Ulgiati (2001) calculated the appropriate economic scale of a region under the constraints of resource and environmental carrying capacity of the United States (US) based on energy analysis methods. Wackernagel and Rees (1996) proposed and improved the ecological footprint method to calculate and analyze the ecological footprint and ecological carrying capacity of more than 50 countries worldwide. Costanza et al. (1998) and Martinez et al. (2007) conducted research on the marine ecological economy and analyzed the importance of marine ecology, economy, and society to achieve sustainable development and utilization of marine resources to the maximum value. In addition, Kildow and McIlgorm (2009) believed that there are many problems in the current marine ecology and marine industry, and appropriate economic and ecological recovery measures need to be applied for different marine and coastal areas.

Based on a comprehensive literature review and expert interviews, Jiang, Chen, and Dai (2017) established an evaluation system for the carrying capacity of marine industrial parks by using 32 indexes covering the dimensions of pressure, carrying capacity, and transformation. They comprehensively evaluated the carrying capacity of marine industrial zones in Shandong province by using the state space and analytic hierarchy process methods together. Céline, Agnès, and Patrick (2008) conducted empirical analysis of the carrying capacity of French coastal areas based on the carrying capacity assessment framework and proposed relevant suggestions to the national and local governments in coastal areas facing population growth and important tourism flows. Wang, Wang, and Song (2017) analyzed the marine ecological status and influencing factors of Qingdao and its coastal areas from 1982 to 2015 by establishing a comprehensive identification method, including the selection of potential factors, level of independence test, and identification of controlling factors and water quality effects. With the deepening of research on ecological carrying capacity, scholars began to link ecological carrying capacity with industrial layout structure and discussed the method to optimize regional industrial layout from the perspective of ecological carrying capacity (Cheung & Sumaila, 2008; Henry, Barkley, & Evatt, 2002; Read & Fernandes, 2003; Victoria, 2016). Ma, You, Ma, Xie, and Li (2012) evaluated the sustainability of the marine ecosystem of Dongtou islands by constructing a marine ecological carrying capacity index system based on a general conceptual model. Wang, Zhang, and Chen (2010) analyzed the industrial functional structure of the marine economy in Liaoning province in 1997 and 2006 using principal component analysis and proposed suggestions for constructing a "five-point and one-line" coastal economic zone consistent with the industrial functional structure and layout of the marine economy. Hsieh and Li (2009) explored the practical experience of the development of deep-sea aquaculture clusters in the US, Japan, and Taiwan and analyzed the interactions between the economy, society, and technology of the three countries from the perspective of industrial clusters. Using the spatial econometric model, Hong and Cheng (2016) tested the spatial correlation of 11 coastal provinces and cities in China from 2001 to 2013 by analyzing the factors affecting marine industrial structure upgrading and the various factors influencing industrial structure in two samples with different industrial structure levels.

To sum up, the current index system for evaluating marine ecological carrying capacity has the following limitations: (1) the current index system does not distinguish between pressure bearer and pressure giver in a marine ecological carrying capacity system but consider them as a whole and overlooks the fact that the two sides are relatively independent; for example, on the one hand, the amount of marine resources reserves reflects the carrying capacity of the ocean, and on the other hand, it reflects the strength of human demand for marine resources; and (2) in determining the hierarchy of marine ecological carrying capacity, most scholars choose an isometric partition with certain subjectivity and most of the literature on the evaluation of marine ecological carrying capacity indexes cannot determine the current state of marine ecological carrying capacity accurately: surplus, equilibrium, or overload. Therefore this chapter screens the relevant indexes of marine ecological carrying capacity, considers the marine ecological system as the pressure bearer and various human activities on the marine system as the pressure giver for analysis, and divides the marine ecological carrying capacity into three states, surplus, equilibrium, and overload based on the coupling relationship between the pressure bearer and the pressure giver. Finally, this index system is applied to research on the marine ecological carrying capacity of 11 coastal areas in mainland China seeking a balance point between the marine ecosystem and human activities and provides reference for the sustainable development of China's marine ecosystem.

2.3 Evaluation system and method of marine ecological carrying capacity

2.3.1 Evaluation systems of marine ecological carrying capacity

In recent years, the application of a system dynamics method for evaluating marine ecological carrying capacity indicates that people regard a marine ecosystem as a combination of subsystems such as population, economy, space resources, and environment (Hong & Cheng, 2016). Overall, marine ecological carrying capacity can be divided into marine resources and human activities. Marine resources include marine products, the coastline, and marine utilization areas, which is the carrier of marine ecological carrying capacity; the system of human activities refers to economic growth and social development through utilizing marine resources and a series of activities carried out by people against the marine ecological environment, namely, the pressure giver.

The theory of man−earth relationship is the theoretical base for studying marine ecological carrying capacity. This theory is the main body and core of geographical research. Human production activities have positive or negative effects on the environment, while the geographical environment also limits human survival and production and reacts to human beings or even plays a role in delaying or promoting social development. The world is facing the challenges of population expansion, resource depletion, environmental degradation, and other issues such as resource supply−demand imbalance in time and region, sharp depletion of forest reserves, land desertification, and reduction of resources leading to increased demand for ocean resources. Furthermore, acquiring living space from the ocean has become a new target for countries. Presently, human beings, as the pressure giver, obtain survival capital from the ocean through various ways of development and utilization, and the ocean, as the pressure bearer, releases resource supply through their own metabolism. This is a reflection of the profound man−earth relationship. The linkage and coupling relationship between marine resources and human activities is described in Fig. 2.2 (Jin, Jin, & Li, 2018).

Therefore starting from the man−earth relationship theory, this chapter considers the limit of the pressure bearer (marine resources system) to support the pressure giver (human activities system) to represent the marine ecological carrying capacity. Specifically, on the premise of classification of marine resources, this chapter subdivides "pressure-bearing" indexes into "gross," "structure," and "intensity" indexes. By referring to Jiang et al. (2017), Di, Porter, and Tracey (2003), and Lu, Yuan, Lu, and Su (2018) construction of a marine ecological carrying capacity evaluation index system, "pressure-bearing" indexes include mariculture yield, wetland area in coastal areas, sea salt yield, per capita water resources, per capita nearshore and shore, unit area marine industry output, and unit coastline economic output. Pressure giver refers to the subsystem of human activities that is divided into three types, "economy," "society," and "ecology," as human activities are generally present in these three aspects. Specifically, "pressure giving" includes the total output of the main marine industries, fixed asset investments, number of coastal

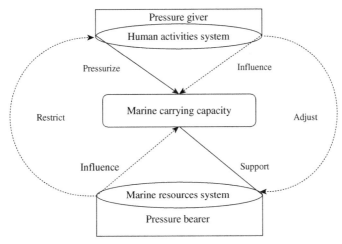

Figure 2.2 Coupling relationship of marine ecological carrying capacity.

pollution treatment projects, number of marine-related employees, and total direct discharge of industrial wastewater into the ocean.

2.3.2 Marine ecological carrying capacity index

Based on completing the construction of marine ecological carrying capacity, the pressure-giving and bearing indexes are calculated by using the gray correlation entropy method, and the marine carrying capacity index is calculated by using the ratio of the two. Besides, the partition of carrying capacity states is accomplished through the pressure giving and bearing coupling curve.

2.3.2.1 Standardization of index

It is not easy to analyze a multiindex system due to the presence of different units, dimensions, and orders of magnitude, and the evaluation results may even be affected. Therefore to unify standards, all evaluation indexes should be standardized to eliminate dimensions and converted into dimensionless standard components with no orders of magnitude for analysis and evaluation. The range transformation method is used to standardize the original data:

$$y_{ij} = \frac{(1 - \Delta) + \Delta \times \left[x_{ij} - \min\left(x_j\right)\right]}{\max\left(x_j\right) - \min\left(x_j\right)} \tag{2.1}$$

$$y_{ij} = \frac{(1 - \Delta) + \Delta \times \left[\max(x_j) - x_{ij}\right]}{\max(x_j) - \min(x_j)} \qquad (2.2)$$

where $i \in [1, m]$, $j \in [1, n]$, and $\Delta = 0.9$; y_{ij} refers to the standard value of the jth index in the ith area, x_{ij} represents the standard value of the jth index in the ith area; $\min(x_j)$ and $\max(x_j)$ represent the minimum value and the maximum value of the jth index in the ith area. In addition, if the corresponding index has a positive influence on the marine ecological carrying capacity, formula (2.1) will be used; otherwise, formula (2.2) will be used.

2.3.2.2 Weight confirmation

Since there is no definite index system for evaluating the marine ecological carrying capacity at present, there is inevitably some uncertainty in the selection of indexes. To ensure the relative objectivity of the results, this chapter adopts objective weight determination methods to calculate the weight of each index, among which the gray correlation entropy method is an objective weighting method that optimizes the traditional gray correlation analysis. The deviations caused by subjective factors can be avoided by properly weighting the influencing factors of the marine ecological carrying capacity, namely, judging the utility values of indexes by using information inherent in the evaluation indexes. The specific steps are as follows:

1. Calculating the correlation coefficient. Such a coefficient reflects the proximity of standardized values to the standardized reference value, and the larger it is the better. The specific formula is as follows:

$$\zeta_{ij} = \frac{\min_i \min_j \left|y'_{0j} - y'_{ij}\right| + \rho \max_i \max_j \left|y'_{0j} - y'_{ij}\right|}{\left|y'_{0j} - y'_{ij}\right| + \rho \max_i \max_j \left|y'_{0j} - y'_{ij}\right|} \qquad (2.3)$$

where y'_{0j} is the reference column that is composed of the optimal values of standardized indexes of each term. Set the reference sequence of the number as $y'_{0j} = [y_{01}, y_{02}, \ldots, y_{0n}]$, among which, when the index is positive, the reference value will be maximum value in all the areas; if negative, then select the minimum value; ρ is the resolution ratio that equals to 0.5; from formulae (2.1) and (2.2), we can know that the value range after standardization is $[0.1, 1]$; hence, $\min_i \min_j \left|y'_{0j} - y'_{ij}\right| = 0$ and $\max_i \max_j \left|y'_{0j} - y'_{ij}\right| = 0.9$; then, formula (2.3) can be simplified to $\zeta_{ij} = 0.45 / \left(1.45 - y'_{ij}\right)$.

2. Solving the gray entropy. Information is an index to measure the degree of system orderliness while entropy is an index to measure the degree of system disorderliness. The less the entropy, the less the uncertainty and greater is the information of index and the bigger the role of the comprehensive evaluation and corresponding weight value. The difference between gray entropy and Shannon entropy is that the latter is a probabilistic entropy with certainty, while gray correlation entropy is a nonprobabilistic entropy with gray uncertainty. The gray correlation entropy of the jth index can be expressed as

$$H_j = -\frac{1}{\ln m}\sum_{i=1}^{m} h_{ij}\ln h_{ij} \tag{2.4}$$

where $h_{ij} = \zeta_{ij}/\sum_{i=1}^{m}\zeta_{ij}$; $i\in[1,m]$, $j\in[1,n]$; $h_{ij}\geq 0$, $\sum_{i=1}^{m}h_{ij}=1$.

3. Confirming weight. The degree of deviation of the jth index is $\varepsilon_j = 1 - H_j$ and the weight coefficient is

$$W_j^L = \frac{\varepsilon_j}{\sum_{j=1}^{n}\varepsilon_j} \tag{2.5}$$

2.3.2.3 Computational formula of marine ecological carrying capacity index

In this chapter, the linear weighting method is adopted to calculate the pressure-giving and bearing indexes for which the computational formula is as follows:

$$S_i = \sum_{j=1}^{n} S_{ij}' W_j^S \tag{2.6}$$

$$P_i = \sum_{j=1}^{n} P_{ij}' W_j^P \tag{2.7}$$

Finally, the marine ecological carrying capacity can be expressed as $C_i = S_i/P_i$.

2.3.3 Measurement method of marine industrial structure

When calculating the industrial structure system, we can divide the system into n industrial categories; h_i represents the industrial level of the ith industry and k_i represents the proportion of output of the ith industry to total output. Then, computational formula H of the industrial structure level of this system can be expressed as

$$H = \sum_{i=1}^{n} k_i h_i \tag{2.8}$$

In this chapter, we divide the marine industry structure into primary, secondary, and tertiary systems. Therefore the value of n is 3. As the development level of an industry can be expressed by the labor productivity corresponding to the industry, this chapter adopts labor productivity to measure the industrial level of a certain industry in the system. Labor productivity is represented by the ratio of the output value of a certain industry (p_i) to the number of employees in the industry (l_i). Therefore the industrial structure level of a certain system H can be expressed as

$$H = \sum_{i=1}^{n} k_i \times \frac{p_i}{l_i} \tag{2.9}$$

Besides, due to the big difference in labor productivity between different industries, it is not conducive to observe the changes i enterprises with low labor productivity. Therefore to improve the sensitivity of changes in the level of industrial structure, this chapter deals with formula (2.9) as follows without essential influence on the variation trend and relative size of various industrial structures:

$$H = \sum_{i=1}^{n} k_i \times \sqrt{\frac{p_i}{l_i}} \tag{2.10}$$

where p_i/l_i represents the industrial level of the ith industry, and the larger the value, the higher the industrial level of the corresponding industry; $k_i \cdot \sqrt{p_i/l_i}$ represents the contribution of the ith industry in the optimization process of the industrial structure level, and the larger the value, the greater the contribution.

2.3.4 Partition of pressure giving and bearing coupling curve

Methods to measure coupling between two or more systems can be divided into two: (1) directly calculate the coupling degrees of the system by using the coupling degree formula, and (2) reflect the different coupling degrees of the system by using the coupling curve. The coupling degree model can only be used to reflect the degree of coupling among

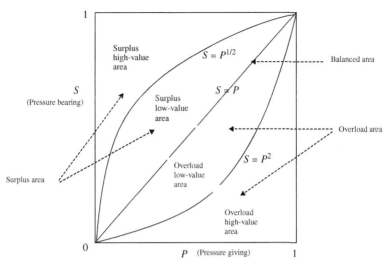

Figure 2.3 Pressure giving and bearing coupling curve.

the subsystems to be studied and it cannot directly classify the coupling degree, while the coupling curve can classify the value of the coupling degree. The coupling curve not only considers the influence of critical values but can also construct suitable mathematical expressions flexibly according to different research objects and content, which is more convenient to use. Therefore in this chapter, the coupling state of marine ecological carrying capacity is divided by using the coupling curve, and its critical value is obtained by mathematical transformation.

As shown in Fig. 2.3, the horizontal axis P represents the pressuring-giving index and the vertical axis S represents the pressure-bearing index. The critical values of marine carrying capacity they reflect are $S = P$, $S = P^2$, and $S = P^{1/2}$, respectively. According to the critical value curve $S = P$, we can divide the marine ecological carrying capacity indexes into three states: $C = 1$, $C > 1$, and $C < 1$. When $C = 1$, it means that the marine ecological system carrying capacity is in equilibrium state, with no surplus or overload; when $C > 1$, it means that the pressure due to human activities has not exceeded the carrying capacity of the marine ecosystem and there is still some room for development and utilization, and it is in a surplus state; when $C < 1$, it means that the marine ecosystem is overloaded. On this basis, the surplus area and the overload area are subdivided according to the critical curves $S = P^2$ and $S = P^{1/2}$: as to overload area $(C < 1)$, when $S < P^2$, it means that human activities concerning marine resources have seriously exceeded the carrying capacity of the marine ecosystem and the relationship is in the "overload high-value area;" when $S > P^2$, it means that though human activities have exceeded the limit of the carrying capacity of the marine ecosystem, the situation is not very critical and their relationship is in the "overload low-value area;" as to the surplus area $(C > 1)$, when $S > P^{1/2}$, it means that the marine ecosystem has strong pressure-bearing capacity and development potential, that is, in the "surplus high-value area;" when $S < P^{1/2}$, it means that though the marine ecosystem has not reached its limit of carrying capacity, there is little room for development; the relationship between marine resources system and human activities system is relaxed and in the "surplus low-value area."

2.3.5 Quadrant diagram identification and classification method

The identification and classification method of quadrant diagram can be used as the objective standard for the judgment of matter relationship, and it is a relationship recognition method (Chen, Lu, & Cha, 2009). Therefore this chapter analyzes the relationship between the level of marine industrial structure and marine ecological carrying capacity during the research period by using the corrected quadrant diagram identification method, which includes the degree of deviation. The details are as follows:

First, the calculated marine industrial structure level (T) and marine ecological carrying capacity index (B) are standardized:

$$Z_T = \frac{\left(T_{ij} - \overline{T}\right)}{S_T} \tag{2.11}$$

$$Z_B = \frac{\left(B_{ij} - \overline{B}\right)}{S_B} \tag{2.12}$$

where Z_T and Z_B represent the standardized marine industrial structure level and the marine ecological carrying capacity, respectively; $\overline{T}$ and $\overline{B}$, respectively, represent the average value of the two indexes in the ith year, and S_T and S_B represent the standard deviation of the two indexes in the corresponding years, respectively.

Second, the quadrant diagram of the relationship between the level of marine industrial structure and ecological carrying capacity is constructed, as shown in Fig. 2.4: Z_B is the horizontal coordinate and Z_T is the vertical

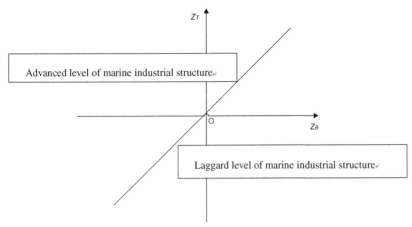

Figure 2.4 Relationship between marine industrial structure level and ecological carrying capacity.

coordinate. Each point in the quadrant represents the level of marine industrial structure and ecological carrying capacity in different regions in different years.

Finally, the relationship between the level of marine industrial structure and ecological carrying capacity is judged. We stipulate that $(Z_T + Z_B)$ represents the degree of coordination C between the marine industrial structure and the ecological carrying capacity, and $|Z_T - Z_B|$ represents the degree of deviation D between the marine industrial structure and the ecological carrying capacity. Subsequently, the types and classification criteria of industrial structure, degree of coordination and degree of deviation are formulated, and the results are shown in Table 2.1. Finally, the relationship types between marine industrial structure and ecological carrying capacity are classified.

2.4 Empirical analysis of marine ecological carrying capacity

2.4.1 Sample selection and data processing

By considering the 11 coastal provinces and cities (Liaoning, Hebei, Shandong, Jiangsu, Zhejiang, Fujian, Guangdong, Hainan, Tianjin, Shanghai, and the Guangxi Zhuang autonomous region) as the research

Table 2.1 Types and standards of evaluation parameters of marine industrial structure level and ecological carrying capacity.

Evaluation parameters	Types and classification standards		
Marine industrial structure level	Low level (III) $0.5C \leq 0$	Medium level (II) $0 < 0.5C \leq 0.5$	High level (I) $0.5C > 0.5$
Type of coordination	Laggard industrial structure $Z_T < Z_B$		Advanced industrial structure $Z_T > Z_B$
Degree of deviation	Mild deviation (c) $0.1 < D \leq 1$	Moderate deviation (b) $1 < D \leq 2$	Severe deviation (a) $D > 2$

Note: When $0 < D \leq 0.1$, it means that marine economic quality and marine economic scale are in a coordinated state.

objects and relying on the train of thought about the construction of an index system in the above text, 10 pressure-bearing indexes and 10 pressure-giving indexes are determined, as shown in Table 2.2. The data for each index are obtained from China Marine Statistical Yearbooks for the respective years.

1. The "pressure-bearing" index is characterized from three aspects: gross, structure, and intensity, representing the carrying capacity of the marine ecosystem. The mariculture yield and the sea fishing yield in the gross dimension are used to reflect the status of water products in the ocean; the wetland area in coastal areas reflects the status of the marine ecological environment; the mariculture area reflects the current scale of mariculture; the sea salt yield reflects the status of the marine product storage. In the structure dimension, the per capita water resource is used to reflect the content of fresh water in the coastal areas; the per capita nearshore and shore and the per capita sea area are used to reflect the status of marine resources after deducting the influence of the population base. In the intensity dimension, the unit area marine industrial output and the unit coastline economic output are used to reflect the comprehensive output capacity of industries and economy in the range of waters with confirmed rights.

2. The "pressure-giving" indexes describe the pressure created by a series of human activities on marine resources from three aspects, society, economy, and ecology. In the economy dimension, the total output of main marine industries and the fixed asset investments reflect the overall economic strength of each coastal area; the total investment in

Table 2.2 Marine ecological carrying capacity evaluation index.

System	Dimension	Index (unit)	Dimension	Index (unit)
Pressure bearer	Gross	Mariculture yield (ton)	Pressure giver	Economy	Total investments in industrial wastewater pollution governance (CNY10000)
		Sea salt yield (10,000 tons)			Total investments in industrial solid wastes pollution governance (CNY10000)
		Wetland area in coastal areas (1000 hectares)			Total output of main marine industries (CNY100 million)
		Sea water breeding area (hectare)			Fixed assets investments (CNY100 million)
	Structure	Sea fishing yield (ton)		Society	Number of coastal pollution treatment projects
		Per capital water resource (m^3/person)			Number of subjects of marine scientific research institutions
		Per capita nearshore and shore (hectare/10,000 persons)			Number of sea-related employees (10,000 persons)
		Per capita sea area (hectare/10,000 persons)			Number of employees in scientific research institutions (person)
	Intensity	Unit area marine industrial output (CNY10000/hectare)		Ecology	Total amount of industrial wastewater discharged directly into the ocean (10,000 tons)
		Unit coastline economic output (CNY10000/1000 m)			Amount of comprehensively utilized industrial solid wastes (10,000 tons)

industrial wastewater pollution governance and the total investments in industrial solid wastes pollution governance reflect the force of pollution governance in coastal areas. In the society dimension, the number of sea-related employees reflects the scale of sea-related employed population; the number of coastal pollution treatment projects, the number of subjects of marine scientific research institutions, and the number of employees in scientific research institutions reflect the degree of completeness in social marine environment improvement and scientific research construction. In the ecology dimension, the total amount of industrial wastewater that is discharged directly into the ocean and the amount of comprehensively utilized industrial solid wastes reflect the intensity of environmental pollution.

2.4.2 Current status of marine ecological carrying capacity

Based on the above theoretical analysis, this chapter uses the relevant data of 11 coastal areas in China from 2006 to 2015 to calculate the pressure-bearing indexes, pressure-giving indexes, and marine ecological carrying capacity index of each area (Table 2.3).

Table 2.3 shows that only a small number of samples (31) had stronger pressure-bearing capacity than pressure-giving capacity, and the overall situation of marine ecological carrying capacity needed to be improved. Among the 11 coastal areas, the pressure-giving indexes of Tianjin, Hebei, Shanghai, Jiangsu, Zhejiang, Shandong, and Guangdong were basically higher than their pressure-bearing indexes, indicating that the social and economic activities in these areas applied heavy pressure on the marine ecological environment and rendered their marine ecological carrying capacity relatively low in all the sample data, where the overall state of marine ecological carrying capacity is disturbing. The marine ecological pressure-bearing indexes and pressure-giving indexes of Shandong province, which is located on the eastern coastal region and the downstream of the Yellow River in China, are always ranked high and closely relate to its total marine resources, development intensity, gross population, and economic output. However, compared with Guangdong province, Shandong province had a higher pressure-bearing index, but the pressure-giving indexes of the two provinces did not differ much. This matches with the fact that Guangdong province has more advanced social and economic activities, indicating that Guangdong has heavier marine ecological pressure and the marine ecological conditions need to be improved.

Table 2.3 Status of marine ecological carrying capacity in 11 coastal provinces and cities from 2006 to 2015.

Year / Area	2006	2007	2008	2009	2010	2011	2012	2013	2014	2015
Tianjin	0.2484	0.2757	0.2807	0.2942	0.3150	0.2894	0.3493	0.4068	0.4037	0.3806
Hebei	0.6932	0.6110	0.5308	0.5322	0.4953	0.4525	0.4666	0.4198	0.4487	0.4469
Liaoning	1.7868	1.4635	1.5663	1.5027	1.2882	1.1965	1.4204	1.418	1.3196	1.2155
Shanghai	0.4801	0.4677	0.7897	0.4626	0.4859	0.4886	0.5710	0.5312	0.4292	0.4467
Jiangsu	0.8429	0.8027	0.6068	0.6663	0.5663	0.5195	0.5641	0.5049	0.5147	0.4692
Zhejiang	1.0500	0.7423	0.9050	0.8653	0.9638	0.7820	0.9165	0.7569	0.7329	0.7593
Fujian	1.2821	1.0703	1.2504	0.9863	0.8997	1.1208	0.9675	0.8561	0.9069	0.8566
Shandong	0.9791	0.9645	0.8483	0.8915	0.9104	0.7240	0.7882	0.9314	0.9333	0.8148
Guangdong	0.6410	0.5941	0.6158	0.5597	0.5565	0.5311	0.5622	0.5957	0.5190	0.4508
Guangxi	1.4330	1.1412	1.3172	1.0598	0.9360	0.9588	1.0174	1.0284	0.8679	0.8908
Hainan	2.0801	2.0466	2.4324	2.5318	2.4319	2.2407	2.0370	2.4136	2.1714	1.8547

Although Tianjin city is consistently under the overload status during the research period, the gap between its pressure-bearing indexes and pressure-giving indexes was declining, connoting that the marine ecological carrying capacity in Tianjin city was improving gradually. The pressure-bearing indexes and giving indexes of Hebei province were both small and the possible reason may be that the scale and intensity of development of marine industries in this area were both relatively low and the marine economy was not developed.

The pressure-bearing indexes of Liaoning, the Guangxi Zhuang autonomous region, and Hainan are slightly higher than their corresponding pressure-giving indexes. Among them, the pressure-bearing indexes of Hainan province during 2006 and 2015 were much higher than its pressure-giving indexes every year, namely, the marine ecosystem in Hainan suffered light pressure and the marine ecological carrying conditions were good overall. This may have resulted from the fact that Hainan owns China's largest waters and is rich in marine resources, which provide natural advantages for local marine industrial development. However, compared with other provinces, the marine economy in Hainan developed relatively slow, for example, the total marine output in 2011 was only CNY61.2 billion, lagging coastal areas such as Guangdong province (CNY980.7 billion), Shandong province (CNY830 billion), Zhejiang province (CNY450 billion), and Fujian province (CNY442 billion). Although the pressure-bearing indexes of Liaoning and Guangxi were slightly higher than their pressure-giving indexes, overall, both the ratios declined during 2006 and 2015, especially the Guangxi Zhuang autonomous region. In 2006 the pressure-bearing indexes were 1.43 times as high as the pressure-giving indexes in Guangxi; however, in 2009 the pressure-bearing indexes were only 1.06 times as high as the pressure-giving indexes, and later, the pressure-bearing indexes were slightly smaller than the pressure-giving indexes. This indicates that with increasing intensity of marine resource development and large-scope human activities, the marine ecological carrying capacity in the Guangxi Zhuang autonomous region was declining continuously.

2.4.3 Provincial dynamic comparison and analysis

Taking 2006 and 2015 as examples, the changes in marine ecological carrying capacity of coastal areas are studied. It can be seen that in 2006 among the 11 coastal areas, 5 had marine ecological carrying capacity

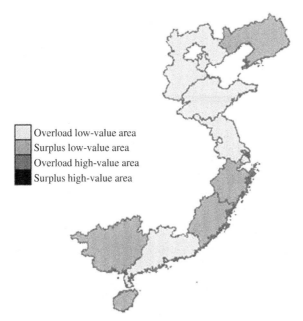

Figure 2.5 Marine ecological carrying capacity of each area in 2006.

index greater than 1 and belonged to the surplus region, namely, Liaoning province, Zhejiang province, Fujian province, the Guangxi Zhuang autonomous region, and Hainan province. Similarly, six areas had marine ecological carrying capacity of less than 1 and belonged to the overload region. By 2012 only Liaoning and Hainan provinces remained in the surplus area with the marine ecological carrying capacity of greater than 1 and both in the surplus low-value area. The other areas belonged to the overload area and the distribution is shown in Figs. 2.5 and 2.6.

By comparing these two years, provinces with changes in marine ecological carrying capacity were mainly Zhejiang province, Fujian province, Guangdong province, and the Guangxi Zhuang autonomous region. Among them, Zhejiang province, Fujian province, and the Guangxi Zhuang autonomous region shifted from the surplus low-value area to the overload low-value area. The ratio of pressure bearing to pressure giving in Zhejiang province decreased from 1.05 to 0.76. Though both the pressure-bearing indexes and the pressure-giving indexes of Zhejiang province increased to a certain extent from 2006 to 2015, the rate of increase for the pressure-bearing indexes was only 5.53% while that for pressure-giving indexes was close to 35%. This indicates that the

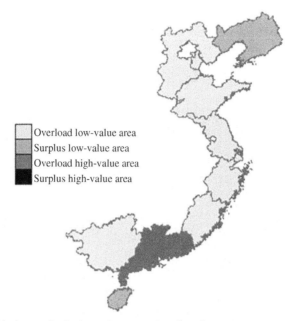

Figure 2.6 Marine ecological carrying capacity of each area in 2015.

protection of marine ecological resources in this region was not enough, and the marine ecological carrying capacity weakened as compared with the past. In the Guangxi Zhuang autonomous region, development and changes were similar to those in Zhejiang province: marine ecological carrying capacity decreased due to the different rates of increase in both the pressure-bearing and giving indexes. During the increase in pressure-giving indexes in Fujian province, the indexes declined slightly as compared with 2006, making the ratio of pressure bearing to pressure giving decrease from 1.282 to 0.857, indicating that marine ecology in Fujian province had further deteriorated under the condition of enhancing marine development. The marine carrying capacity of Guangdong province deteriorated continuously from 2006 to 2015 and shifted from the overload low-value area to the overload high-value area. Therefore when developing the marine economy, Guangdong province should pay special attention to the recovery of marine ecological carrying capacity, slow down the development pace of marine economy, and combine utilization with cultivation for marine ecological environment.

In addition, the marine ecological carrying capacity areas of Tianjin city, Shanghai city, Hebei, Shandong, Liaoning, Jiangsu, and Hainan provinces remained unchanged, but the C value of each area changed to some extent. The biggest change happened in Liaoning province (from 1.787 to 1.215), indicating that though the strength of development and utilization of marine resources increased during 2006 and 2015, protection measures for the marine ecological carrying capacity played a role and there was still some room for development and utilization of the marine ecosystem on the premise of focusing on the degrees of pressure bearing and pressure giving.

2.5 Study on the relationship between marine ecological carrying capacity and industrial structure

2.5.1 Development trends of marine ecological carrying capacity and industrial structure

2.5.1.1 Development level of marine industrial structure

Marine industries are classified as primary, secondary, and tertiary and the level of marine industrial structure is calculated according to the methods mentioned above. The variation trends in marine industrial structure from 2006 to 2015 are obtained, as shown in Fig. 2.7.

From 2006 to 2015 the marine industrial structure of China's coastal provinces showed a growing trend, although with a certain volatility. The level of Shanghai's marine industrial structure ranked first during the research period, while Jiangsu rose from the sixth place in 2006 to the second place in 2015. This indicates that the marine industrial structure in Jiangsu province was optimized with continuous improvement of the

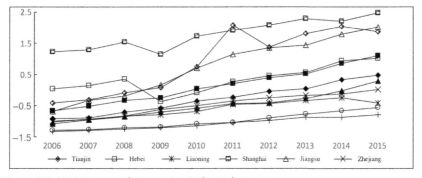

Figure 2.7 Variation trends in marine industrial structure.

management system and development of scientific innovation. The position of Hebei province dropped from second in 2006 to the fifth in 2015. The marine industrial structure levels in Fujian, Guangdong, and Shandong provinces showed a steadily rising trend, with Shandong presenting the largest increase in range, followed by Guangdong and Fujian. The marine industrial structure levels in Hainan and Guangxi provinces ascended slightly but were still low. Until 2015 the marine industrial structure levels in Tianjin, Jiangsu, Hebei, and Zhejiang had gradually surpassed the average marine industrial structure level of coastal areas, while those in Hainan, Guangxi, and Liaoning were still low, compared with the average marine industrial structure level and needed further optimization.

2.5.1.2 Current status of marine ecological carrying capacity

According to the marine ecological carrying capacity calculated by the method of gray correlation entropy, the variation trend chart of marine ecological carrying capacity from 2006 to 2015 is presented in Fig. 2.8.

Except for some provinces such as Liaoning, Hainan, and Guangxi, fluctuations in marine ecological carrying capacity in most of China's coastal areas were minor. Although Guangdong, Tianjin, and Shanghai had obvious advantages in marine economic locations, their marine ecological carrying capacity was at a low level due to excessive pursuit for

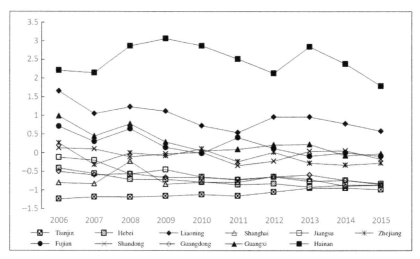

Figure 2.8 Variation trend chart of marine ecological carrying capacity.

rapid economic development and significant intensity in the exploitation and utilization of marine resources.

The marine ecological carrying capacity of Shandong and Zhejiang was at the same level from 2006 to 2015. Hainan province showed an uptrend from 2006 to 2009; however, after 2009 due to the increase in marine resource development intensity and strong homogenous industry competition, the marine ecological carrying capacity decreased, but the ecological environment remained in a good condition. The marine ecological carrying capacity of Hebei, Guangdong, Tianjin, Jiangsu, and Shanghai ranked low as compared with other coastal areas and that of Hebei province was on a small declining trend during the research period.

In addition, the standardized value of marine ecological carrying capacity in Hebei province, Guangdong province, Shanghai city, Tianjin city, and Jiangsu province in 2006 was less than 0, indicating that the marine ecological carrying capacity in these areas was lower than the average level during the research period. In 2015 the marine ecological carrying capacity for only Liaoning province and Hainan province was greater than zero, that is, compared with 2006, the marine ecological carrying capacity in Zhejiang, Fujian, Shandong, and the Guangxi Zhuang autonomous region declined from above average to below average level, indicating that the marine development intensity in these areas was relatively high and the marine ecological protection there should be noted.

2.5.2 Coordinated development of marine ecological carrying capacity and industrial structure

2.5.2.1 Temporal evolution of the relationship between marine industrial structure and marine ecological carrying capacity

According to the levels of marine industrial structure and marine ecological carrying capacity indexes, the change in relationship between the marine industrial structure level and marine ecological carrying capacity of the 11 coastal areas in China from 2006 to 2015 is calculated by using the quadrant diagram identification method and the results are shown in Table 2.4.

The marine industrial structure and carrying capacity of China's coastal areas were seldom in a coordinated state during the research period. In addition, from 2006 to 2009 the marine industrial structure remained mainly at the low level and in laggard status; after 2009, however, the marine industrial structure changed from laggard to advanced status.

Table 2.4 Relationship between marine industrial structure and marine ecological carrying capacity in China's coastal areas.

Area	2006	2007	2008	2009	2010	2011	2012	2013	2014	2015
Tianjin	IIIAa	IIIAa	IIIAb	IIIAb	IIIAb	IIAc	IIAc	IIAc	IAc	IIAc
Hebei	IIIAa	IIIAa	IIIAb	IIIAa	IIIAa	IIIAb	IIIAb	IIIAb	IIAb	IIAb
Liaoning	IICc	IICb	IICc	IICb	IICb	IICa	IICb	IICb	IICb	IICa
Shanghai	IIAc	IICc	IAb	IIAb	IIAc	IIAc	IAc	IAc	IAc	IAc
Jiangsu	IIICa	IIICa	IIIAa	IIIAa	IIAb	IIAb	IIAb	IIAc	IAc	IAc
Zhejiang	IIICb	IIICa	IIICa	IIICa	IIICa	IIICa	IIICa	IIIAa	IIIAa	IIIAa
Fujian	IIICb	IIICb	IIICb	IIICa	IIICa	IIICa	IIICa	IIICa	IIIB	IIIAa
Shandong	IIICa	IIICa	IIICa	IIICa	IIB	IIAa	IIAa	IIAa	IIAa	IIAb
Guangdong	IIICa	IIICa	IIICa	IIIB	IIIAa	IIIAa	IIIAa	IIIAa	IIIAb	IIIAb
Guangxi	IIICc	IIICb	IIICb	IIICb	IIICb	IIICb	IIICb	IIICa	IIICa	IIICa
Hainan	IICc	IICc	ICc	ICc	ICc	ICc	ICc	ICc	ICc	ICc

Note: I, II, and III represent the levels of development in the marine industry such as high, medium, and low, respectively; A, B, and C indicate that the relationship between the scale of marine industry and its carrying capacity is in advanced, coordinated, and laggard status, respectively; a, b, and c represent deviation degrees such as severe, moderate, and mild, respectively.

In Tianjin and Hebei provinces, the marine industrial structure remained in a low and advanced status continually, but the deviation between the marine industrial structure level and the marine ecological carrying capacity was increasing continuously: deviation in Tianjin city changed from moderate in 2006 to severe in 2015, while that in Hebei province changed from mild in 2006 to moderate in 2015, indicating the disparity in marine industrial structure and marine ecological carrying capacity in these two areas, and protection of marine ecology should be strengthened during the optimization of marine industrial structure to improve marine ecological carrying capacity. The marine industrial structure in Liaoning province was at medium development level with a relatively laggard status, implying that attention should be paid to optimize and regulate the marine industrial structure in Liaoning province, while maintaining the current status of marine carrying capacity level. The industrial structure levels in Shanghai and Jiangsu were increasing continuously during the whole research period and were in advanced status after 2008. Therefore these two areas should focus on protecting the marine environment to realize the coordination between industrial structure and ecological carrying capacity at a high level. Industrial structures in Zhejiang, Fujian, Shandong, and Guangdong gradually developed from a laggard to advanced status while those in Guangxi and Hainan were always at the laggard status, especially in Hainan where the industrial structure level severely lagged the marine ecological carrying capacity. The reasons may be due to two aspects: first, marine ecological carrying capacity was too high and marine resources were not fully and efficiently utilized; second, marine industrial structure in Hainan was unreasonable with an excessive proportion of the primary industry. From the variation trend in marine industrial structure in China's coastal areas, we can identify that the coordinated relationship between marine industrial structure level and marine ecological carrying capacity changed with time from low level to high level and laggard status to advanced status. This indicates that people's attention to marine ecological carrying capacity was insufficient when developing the marine industry and there was excessive exploitation of marine resources for marine industrial development.

2.5.2.2 Spatial pattern of coordinated development between marine industrial structure and marine ecological carrying capacity

Relevant data of marine industrial structure and ecological carrying capacity in China's coastal areas in 2006, 2009, 2012, and 2015 are selected to

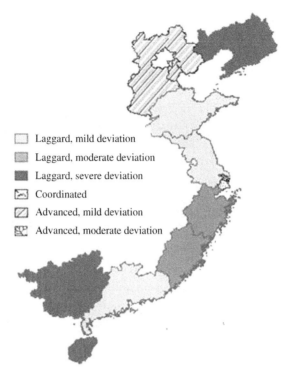

Figure 2.9 Year 2006.

analyze the variations in spatial distribution and the results are shown in Figs. 2.9–2.12.

The relationship between marine industrial structure and marine ecological carrying capacity in China's coastal areas presented the following characteristics in 2016: among the 11 coastal areas, industrial structures in Tianjin, Hebei, and Shanghai were in the advanced status, with Hebei and Tianjin in the mild advanced status and Shanghai in the severe advanced status. The industrial structures in Liaoning, Jiangsu, Zhejiang, Fujian, Shandong, Guangdong, Guangxi, and Hainan were all in the laggard status, with Jiangsu, Shandong, and Guangdong in the mild laggard status, while Zhejiang and Fujian remained in the moderate laggard status, and Liaoning, Guangxi, and Hainan were in a severely laggard status. Broadly, the number of areas with laggard marine industrial structures was higher than that of areas with advanced structure. Until 2009 except in Hebei, Shandong, and Hainan, marine industrial structures in Tianjin, Liaoning, Shanghai, Jiangsu, Zhejiang, Fujian, Guangdong, and Guangxi

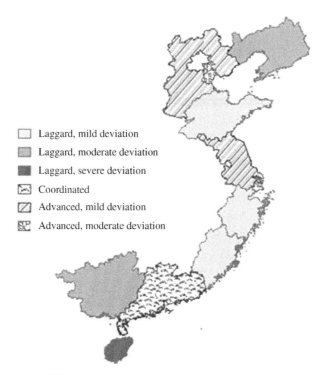

☐ Laggard, mild deviation
▨ Laggard, moderate deviation
■ Laggard, severe deviation
▨ Coordinated
▨ Advanced, mild deviation
▨ Advanced, moderate deviation

Figure 2.10 Year 2009.

had all changed: the structure in Tianjin changed from mild advanced status to moderate advanced status, and the deviation between industrial structure and carrying capacity increased even though the marine ecological carrying capacity in Tianjin presented a slightly upward trend, indicating that the marine industrial structure level in Tianjin had improved continuously and with a greater increase in range. The industrial structures in Liaoning and Guangxi changed from severe laggard status to moderate laggard status, indicating that the gap between marine industrial structure and marine ecological carrying capacity was decreasing and the main reason for this was the optimization of industrial structure; industrial structures in Jiangsu and Guangdong changed from mild laggard status to mild advanced status and coordinate status, respectively. The industrial structures in Zhejiang and Fujian changed from moderate laggard status to mild laggard status, indicating that the degree of deviation between marine industrial structure and ecological carrying capacity was decreasing continuously, which was beneficial for their coordinated development.

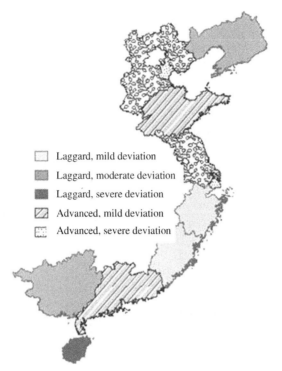

Figure 2.11 Year 2012.

Later, with China's increasing focus on the marine economy, to achieve sustainable development of the marine economy, the marine industrial structure was further optimized, and the development level of marine industrial structure in various coastal areas showed a leading trend gradually. By 2012 the development of marine industrial structure in six areas had reached an advanced level, and the number of areas with advanced industrial structure slightly exceeded the number of areas with lagged industrial structure. The marine industrial structures of Shandong and Guangdong gradually changed from the laggard status to the coordinated status and finally reached a slightly advanced status. The degree of deviation between the industrial structure and the marine ecological carrying capacity in Tianjin, Hebei, Shanghai, and Jiangsu increased, while the status in Liaoning, Zhejiang, Guangxi, Hainan, and Fujian remained unchanged. By 2015 except in Shanghai, Guangxi, and Hainan, marine industrial structures in the other eight areas were all in an advanced status. Overall, the spatial distribution characteristics of China's marine industrial

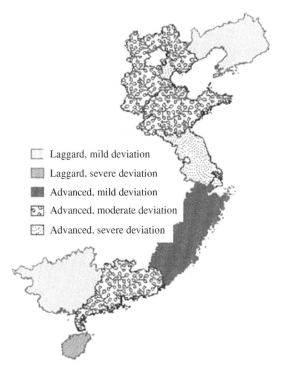

Laggard, mild deviation
Laggard, severe deviation
Advanced, mild deviation
Advanced, moderate deviation
Advanced, severe deviation

Figure 2.12 Year 2015.

structure was as follows: structures in Tianjin, Shanghai, Jiangsu, Hebei, Shandong, and Guangdong were in moderate advanced status; structures in Zhejiang and Fujian were in the mild advanced status; structures in Liaoning and Guangxi were in the mild laggard status, and the structure in Hainan remained in the severe laggard status.

A general survey of the coordinated development of marine industrial structure and ecological carrying capacity in China's coastal areas shows that the main reasons for the development changes include the following aspects. Tianjin, Hebei, and Shanghai have always had an advanced industrial structure due to the relatively reasonable marine industrial structure and low marine ecological carrying capacity. Among all the coastal areas, the marine ecological carrying capacity in these three areas is the lowest, and the industrial structure in Hebei has always been at a low level. The ecological carrying capacity of Liaoning is slightly higher than the development level of its marine industrial structure, thereby lagging its marine industrial structure slightly. The reason may be that in recent years, with the enhancement in scientific research and innovation and improvement

in the management system in Liaoning, marine resources have been fully
and reasonably utilized, and the marine economy has developed rapidly.
Although marine resource endowment in Guangxi and Hainan is relatively high, their respective marine industrial structures continued to be at
a low level compared with the other areas due to the low level of economic development, small population size, weak scientific power, and
lack of marine management system. The marine industrial structures in
Jiangsu, Zhejiang, Fujian, Shandong, and Guangdong changed from the
laggard to advanced status because their structures have been optimized
continuously with constant development and utilization of marine
resources.

2.6 Conclusion and suggestions

2.6.1 Conclusion

Empirical research on China's 11 coastal areas from 2006 to 2015 indicates
that:

1. Overall, the conditions of marine ecological carrying capacity
   improved in the coastal areas. Among all the samples, pressure-bearing
   capacity was stronger for only 31 samples. Social economic activities in
   Tianjin, Hebei, Shanghai, Jiangsu, Shandong, and Guangdong generated relatively high pressure on the marine ecological environment,
   and the marine ecological carrying capacity in these areas was not
   good. However, marine ecological carrying capacity was reasonably
   strong in Liaoning, Guangxi, and Hainan.
2. Judging from the results of partition, there were 3 surplus low-value
   areas and 8 overload low-value areas in 11 coastal areas and no surplus
   high-value area or overload high-value area. Though Tianjin and
   Shanghai were in the overload low-value area, the marine ecological
   pressure-bearing indexes of these provinces were the lowest among all,
   especially that of Tianjin was close enough to reach the overload
   high-value area. Therefore the intensity of marine development
   should be strictly controlled to use the marine resources effectively.
   The marine ecological carrying capacity indexes were the highest in
   Hainan, indicating that there was still room for development and utilization. Hebei, Fujian, Jiangsu, Zhejiang, and Shandong were also in

the overload area, indicating that they should enhance the conservation and protection of marine resources and avoid blind exploitation of marine resources for local economic growth. Liaoning, Guangxi, and Hainan were in the surplus low-value area, where the development intensity should be properly controlled as per actual conditions.

3. Basically, the marine industrial structure level of each area showed a rising trend, gradually improving from laggard to advanced status. The marine ecological carrying capacity was generally at a low level with small changes. The increase in the range of marine industrial structure level was larger than the marine ecological carrying capacity. However, the marine industrial structure of some areas during 2006 and 2015 was always at the low level. For example, though the regional industrial structure in Hebei, Zhejiang, Fujian, Guangdong, and Guangxi reached an advanced status at a low level, it should be further improved with the ecological carrying capacity.

4. Spatial differences in the coordinated relationship between marine industrial structure level and marine ecological carrying capacity in every area were significant and changed constantly during the research period. Until 2015 the marine industrial structure in most areas was in an advanced status and the deviation degrees were high, indicating that the optimization of marine industrial structure and the current carrying capacity were strongly inharmonious. Therefore when optimizing the marine industrial structure, the marine ecological carrying capacity should be further strengthened to improve coordination in China's coastal areas.

2.6.2 Suggestions

Primarily, regulation and control of the marine ecological carrying capacity include pressure reduction, ecological restoration, and improvement of laws and regulations. The development of the marine industry should include the development of marine resources and marine ecological environment and optimize the marine industrial structure within the sustainable range of the marine system, so as to realize a balanced and coordinated development of the marine industrial structure and the marine ecological carrying capacity.

1. Strictly control marine pollution and encourage enterprises to conserve energy and reduce emissions. Under the government's publicity and guidance, it is necessary to urge the public to protect the marine

ecological environment and support the government and society to increase investments in improving the marine ecological environment and restoring it to the original level. Areas in the overload status including Tianjin, Hebei, Shanghai, Jiangsu, Shandong, and Guangdong where the marine ecological environment was under great pressure from the outside, the amount of pollutants discharged into the ocean should be strictly controlled thereafter.

2. Enhance the protection of marine ecological environment and the construction of relevant infrastructure in coastal areas. The marine ecological protection work in China's coastal areas is still insufficient. For example, the total investments in industrial solid wastes pollution governance in Zhejiang's coastal areas were reduced from CNY167.562 million in 2009 to CNY27.46 million in 2010. Therefore it is necessary to emphasize sustainable development of the marine economy.

3. Optimize the industrial structure and raise the scientific and technological level of marine industries. Overall, the scientific and technological level of China's marine industry as well as the degree of industrial clustering is low, the ability of independent innovation is relatively weak, and the number of emerging industries is small. Therefore it is necessary to improve the utilization efficiency of marine resources to further improve the carrying capacity of the sea areas. The government should further promote the transformation and upgrading of marine fisheries, expand and strengthen the shipbuilding and marine engineering equipment manufacturing industries, and vigorously develop marine tertiary industries such as coastal tourism and marine transportation as well as emerging marine industries. In addition, every coastal area should actively establish scientific and technological information exchange platforms and strengthen cooperation in the industry—university—institute coordinate to promote the transfer scientific and technological achievements and share beneficial resources such as information technology.

4. Improve the overall economic performance of the marine industries and narrow the differences between the marine industrial structure level and the marine ecological carrying capacity. Adjust measures to local conditions based on the relationship between the marine economic development level and the marine carrying capacity of each area and formulate development strategies with regional features to stimulate the overall efficiency of the marine system.

References

Brown, M. T., & Ulgiati, S. (2001). Energy measures of carrying capacity to evaluate economic investments. *Population and Environment, 22*(5), 471−501.

Céline, C., Agnès, P., & Patrick, P. (2008). Assessing carrying capacities of coastal areas in France. *Journal of Coastal Conservation, 12*(1), 27−34.

Chen, M. X., Lu, D. D., & Cha, L. S. (2009). Urbanization and economic development in China: An international comparison based on quadrant map approach. *Geographical Research, 28*(02), 464−474.

Cheung, W., & Sumaila, R. U. (2008). Trade-offs between conservation and socio-economic objectives in managing a tropical marine ecosystem. *Ecological Economics, 66* (1), 193−210.

Costanza, R., d'Arge, R., Groot, R., Farber, S., Grasso, M., Hannon, B., ... Belt, M. V. D. (1998). The value of the world's ecosystem services and natural capital. *Ecological Economics, 25*(1), 3−15.

Di, J., Porter, H., & Tracey, M. D. (2003). Linking economic and ecological models for a marine ecosystem. *Ecological Economics, 46*(3), 367−385.

Henry, M. S., Barkley, D. L., & Evatt, M. G. (2002). *The contribution of the coast to the South Carolina economy.* Clemson, SC: Clemson University, Regional Economic Development Research Laboratory (REDRL).

Holling, C. S. (1973). Resilience and stability of ecological systems. *Annual Review of Ecology and Systematics, 4*(4), 1−23.

Hong, A. M., & Cheng, C. C. (2016). The study on affecting factors of regional marine industrial structure upgrading. *International Journal of System Assurance Engineering and Management, 7*(2), 213−219.

Hsieh, P., & Li, Y. R. (2009). A cluster perspective of the development of the deep ocean water Industry. *Ocean & Coastal Management, 52*(6), 287−293.

Lu, Y., Yuan, J., Lu, X., & Su, C. (2018). Major threats of pollution and climate change to global coastal ecosystems and enhanced management for sustainability. *Environmental Pollution, 239*, 670−680.

Jiang, D. K., Chen, Z., & Dai, G. L. (2017). Evaluation of the carrying capacity of marine industrial parks: A case study in China. *Marine Policy, 77*, 111−119.

Jin, Y. Y., Jin, X. M., & Li, C. (2018). Applying supporting-pressuring coupling curve to the evaluation of urban land carrying capacity: The case study of 32 cities in Zhejiang province. *Geographical Research, 37*(06), 1087−1099.

Kildow, J. T., & McIlgorm, A. (2009). The importance of estimating the contribution of the oceans to national economies. *Marine Policy, 34*(3), 367−374.

Ma, C. H., You, K., Ma, W. W., Xie, J., & Li, F. Q. (2012). A study on marine region carrying capacity and eco-compensation. *Journal of Ocean University of China (English Edition), 11*(2), 253−256.

Martinez, M. L., Intralawan, A., Vázquez, G., Pérez, M. O., Sutton, P., & Landgrave, R. (2007). The coasts of our world: Ecological, economic and social importance. *Ecological Economics, 63*(2−3), 254−272.

Park, R. E., & Burgess, E. W. (1920). Introduction to the science of sociology. *The University of Chicago Press, 131*(6), 1−12.

Read, P., & Fernandes, T. (2003). Management of environmental impacts of marine aquaculture in Europe. *Aquaculture, 226*(1), 139−163.

Slesser, M. (1990). *Enhancement of carrying capacity option ECCO* (pp. 86−99). London: The Resource Use Institute.

Victoria, S. (2016). Mobility and emplacement in north coast Papua New Guinea: Worlding the Pacific Marine Industrial Zone. *The Australian Journal of Anthropology, 27* (1), 30−48.

Wackernagel, M., & Rees, W. E. (1996). *Our ecological footprint: Reducing human impact on the earth*. Gabriola Island: New Society Publishers.

Wang, D. L. (2013). Research on the relationship between the change of marine industry structure and the growth of marine economy in Fujian province. *Ocean Development and Management, 30*(09), 85–90.

Wang, D., Zhang, Y. G., & Chen, S. (2010). Study on the evolution pattern of marine industry functional structure and spatial structure in Liaoning province. *Economic Geography, 30*(3), 443–448.

Wang, S. H., Wang, Y. C., & Song, M. L. (2017). Construction and analogue simulation of TERE model for measuring marine bearing capacity in Qingdao. *Journal of Cleaner Production, 167*, 1303–1313.

CHAPTER THREE

Analysis of influencing factors and efficiency of marine resource utilization in China

3.1 Introduction

The 21st century represents a new era in marine economic development. In this era, we are faced with gradual shortages of land resources, population growth, and increasingly prominent problems of environmental degradation. As a result, a comprehensive and economical marine economy with dual development of land and sea has inevitably become one of the great powerhouses driving the sustained economic growth of various countries. As a vast area covering 70% of the earth, the ocean has become the last frontier for human development. Hence, it is increasingly valued by various coastal countries and competition for the development of marine resources has gradually increased. Since China promulgated and implemented the "National Marine Economic Development Planning Outline" in 2003, various coastal areas have set off a frenzy of building "strong coastal provinces," which shows that the marine economy is also providing a new direction for China's economic growth.

However, while developing the marine economy, balance should be maintained among marine environment protection and governance, economic development, and social development. It is evident from the developmental processes of the marine economy that people have excessively emphasized "quantity," while ignoring the importance of "quality," causing a major waste of marine resources. This clearly violates China's economic development concept of "being good and fast at the same time" and has greatly affected the sustainable development of China's marine economy. With the advance of the 13th 5-year plan, China's marine economy has reached a new level, but it is still in the primary stage of development, which means that the development of China's marine economy is still immature. This requires that the developmental characteristics of China's marine economy in and after the period of the 13th 5-year plan should be grasped to realize the coordinated development, scientific planning, and sustainable development of land and sea as resources.

In the land economy, the study on resource utilization efficiency has become the focus of the study on sustainable economic development. However, due to the limitations of statistical data and research methods in marine research, there are few studies on the utilization efficiency of marine resources. Therefore based on the real environment of China's marine economic development and the problems faced by and the status quo of similar research, this chapter calculates and systematically analyzes the utilization efficiency of marine resources in China. First, indexes for calculations are determined based on the existing literature, the status quo of China's marine economic development, and accessible data; then, the marine resource utilization efficiency is estimated based on the superefficiency slacks-based measure (SBM) with undesirable output and metafrontier models. In addition, this chapter uses kernel density estimation and Markov chain estimation methods to evaluate the dynamic evolutionary laws of marine resources from two dimensions, and further adopts the symbolic regression method to analyze the influencing factors of marine resource utilization efficiency. Thus a more comprehensive analysis of the status quo in the development of China's marine resources utilization and related problems can be carried out to achieve the sustainable utilization of marine resources, namely the coordinated development among "marine economy, ecological environment, and social development."

3.2 Literature review

3.2.1 Productivity research method

Most studies on productivity use parametric methods, mainly the positive production frontier model (Aigner & Chu, 1968) and the stochastic production frontier model (Meeusen & van Den Broeck, 1977), as well as nonparametric methods. Data envelopment analysis (DEA) is a nonparametric research method that mainly obtains the total factor productivity by constructing the best production frontier and comparing each decision-making unit (DMU) with the "best practitioner."

DEA was first proposed by Charnes, Cooper, and Rhodes (1978) with the so-called CCR (constant returns to scale [CRS]) model. Then Banker, Charnes, and Cooper (1984) put forward the BCC (variable returns to scale [VRS]) model. Since then, DEA has been widely used in the field of efficiency research, and later the radial, angular DEA, and the extended models have gradually come into being (Li, Cao, & Ding et al., 2017; Shwartz, Burgess, & Zhu, 2016; Wanke, Barros, & Nwaogbe, 2016). However, the original DEA methods ignored slack variables, leading to the defect that the estimated energy efficiency values were too high (Fukuyama & Weber, 2009). To solve this problem, Tone (2001) introduced the nonradial SBM into DEA to measure technical efficiency. After this development, many scholars started to focus on using SBM models to measure efficiencies. For example, Tavassoli, Faramarzi, and Saen (2014) used the SBM-NDEA model to analyze the efficiency and effectiveness of airline performance in the case of shared input; Yu (2010) conducted airport performance evaluation based on the SBM-network DEA model; Wu (2010) used the SBM model to evaluate the efficiency of the international nonlife insurance industry; Wang, Li, and Wang (2012) carried out SBM efficiency analysis of the industrial banks of China based on the three-stage DEA model; Yuan, Chen, and Shao (2015) evaluated the efficiency of insurance e-commerce websites based on the SBM–DEA model; and Zhong (2011) conducted a comparative study on efficiency among OECD member countries and China based on an SBM model.

With the deepening of research, people began to consider the influence of undesirable outputs on efficiency. At present, undesirable outputs are mainly included in efficiency evaluation through three following methods: input–output transposition (Hailu, 2003), forward attribute conversion (Seiford & Zhu, 2005), and directional distance function (Färe, Grosskopf, & Pasurka, 2007). However, these methods do not consider the slackness of inputs and outputs, thus there is often a deviation between the evaluation results and the actual situation. In the SBM model proposed by Tone (2003), slack variables are directly incorporated into the target function, which solves the associated problems of how to consider undesirable outputs and how to solve the slackness of inputs and output at the same time. Since then, this method has been widely used in efficiency evaluation (Li, 2014), and the results of calculations and analysis are fairly credible. For example, Zhou, Ang, and Poh (2006) calculated environmental efficiencies considering undesirable outputs using the SBM model.

3.2.2 Current situation of marine resources utilization evaluation

3.2.2.1 Current situation of marine resource utilization efficiency evaluation

In terms of the existing studies, scholars' evaluations of the utilization efficiency of marine resources have provided a lot of inspiration for research on marine resources. However, there are still some defects in this utilization efficiency research, such as small quantities, single methods, and narrow perspectives. A review of previous studies reveals that most research on the utilization efficiency of marine resources has focused on a certain sector or industry of the marine economy, and there is a lack of overall research on the growth of the marine economy (Jamnia, Mazloumzadeh, & Keikha, 2015; Maravelias, Tsitsika, 2008; Tingley, Pascoe, & Coglan, 2005; Tongzon, 2001). In addition, few studies analyzing the utilization efficiency of marine resources consider the regional heterogeneity of production technology among different regions, resulting in particular deviations in the calculated efficiencies (Battese, Rao, & O'Donnell, 2004), while the metafrontier method can solve this problem well (Yu, Choi, 2015; Zhang, Sun, & Tang, 2013). Moreover, as most previous domestic studies have not considered undesirable outputs, this chapter combines the superefficiency SBM-undesirable output model and the metafrontier method (or metafrontier function) to measure the total factor productivity of China's marine resources, taking regional heterogeneity into account, which fills a gap in most of the research on marine production technology in this region. In addition, it is possible to rank efficiency values that are higher than one.

3.2.2.2 Current situation of research using the distributed dynamic method

The distributed dynamic method can be divided into the dispersed type and the continuous type, according to the status of the research sequence, corresponding to the Markov chain method and the kernel density estimation method, respectively. This approach was first put forward by Quah (1993), as he thought that the testing of β convergence did not fully reflect the dynamic evolution of study variables. In 1997 Quah (1997) took the per capita income of 105 countries as a research basis, using the kernel density estimation method for analysis, and successfully obtained the density distribution results. Since then, the distributed dynamic method has been widely used. Shi and Huang (2009) conducted a dynamic analysis of the economic growth of Chinese provinces based on kernel density estimation, while Produit, Lachance, Strano, and Joost (2010) applied the kernel density estimation of the network to the analysis of economic activities in Barcelona.

3.2.2.3 Current situation of research on factors influencing marine resource utilization efficiency

So far, with the rapid development of the marine industry, the efficiency of the marine economy and its influencing factors have gradually attracted the attention of scholars. Many scholars have used tobit models to discuss the factors influencing efficiency (Song, Song, An, & Yu, 2013; Akpoko, 2007; Cui & Li, 2015). The tobit model is also widely used in the field of marine economy. For example, Wang and Gai (2018) analyzed the external factors that affect income efficiency using a tobit regression model, whereas Ding, Zhu, and He (2015) established a tobit model based on panel data to analyze the factors influencing marine economic growth with green total factor productivity in 11 coastal areas of China. However, the tobit model has a defect that cannot be ignored, that is, when using tobit models to analyze the influencing factors, the assumption must be made that there is a predetermined structure among these factors, which is usually set as a linear function. Due to the complexity of nonlinear socioeconomic systems, it is almost impossible for such linear structures to extract the true hidden laws underlying the total factor productivity of the ocean. Different studies that use different regression structures will generate completely different results.

To avoid the defects of the tobit model, this chapter finally selected the symbolic regression method to analyze the influencing factors, which was first proposed by Koza (1992) and is widely used nowadays. Yang, Li, Wang, and Zhang (2016) analyzed the influencing factors of energy intensity in various regions of China using symbolic regression and establishing the corresponding symbolic regression model. Cai, Pacheco-Vega, Sen, and Yang (2006) analyzed heat transfer measurements using the symbolic regression approach. Yang, Han, and Chen (2015) also used symbolic regression when predicting the oil production of various countries in the world. Compared with tobit models, symbolic regression can not only show the important factors, but can also automatically find linear or nonlinear relationships without predetermining a regression structure in studying the influencing factors of marine total factor productivity.

3.3 Evaluation method of marine resource utilization efficiency

3.3.1 SBM-undesirable output model

DEA is a systematic analytical method that considers multiinput and multioutput situations, and makes relative evaluations of the efficiency between factor input and output (Charnes et al., 1978).

Suppose, the marine resource utilization efficiency structure has n DMUs and each DMU contains one input vector and two output vectors, which are desirable output and undesirable output, respectively; $x \in R^m$, $y^e \in R^a$, and $y^n \in R^b$; m, a, and b represent input factors of type m, desirable output of type a, and undesirable output of type b, respectively, that are held by each DMU.

Matrix X refers to $X = [x_1, \ldots, x_n] \in R^{m \times n}$ and Y^e refers to $Y^e = [y_1^e, \ldots, y_n^e] \in R^{a \times n}$; Y^n means $Y^n = [y_1^n, \ldots, y_n^n] \in R^{b \times n}$; and suppose that X, Y^e, and Y^n all exceed 0.

Under the condition of CRS, the production possibility set is defined as:

$$P = \left\{ (x, y^e, y^n) | x \geq X\lambda, y^e \leq Y^e\lambda, y^n \leq Y^n\lambda, \lambda \geq 0| \right\} \quad (3.1)$$

where $\lambda \in R^n$, $\lambda \geq 0$ stands for CRS.

The SBM-undesirable output model can be expressed as:

$$\rho^* = \min \frac{1 - \dfrac{1}{m}\sum_{i=1}^{m} \dfrac{D_i^-}{x_{ik}}}{1 + \dfrac{1}{a+b}\left(\sum_{r=1}^{a} \dfrac{D_r^e}{y_{rk}^e} + \sum_{h=1}^{b} \dfrac{D_h^n}{y_{hk}^n}\right)} \quad (3.2)$$

Subject to $x_k = X\lambda + D^-$

$y_k^e = Y^e\lambda - D^e$

$y_k^n = Y^n\lambda + D^n$

$\lambda \geq 0, \quad D^- \geq 0, \quad D^e \geq 0, \quad D^n \geq 0$

where D^-, D^e, and D^n represent slack variables that respectively stand for excessive input, insufficient desirable output, and excessive undesirable output; $\rho^* \in [0, 1]$ stands for the efficiency of the DMU: when $\rho^* = 1$, $D^- = 0$, $D^e = 0$, the DMU is totally efficient; otherwise, there will be efficiency loss and adjustments can be made to realize the optimal efficiency.

The superefficiency SBM model including undesirable output can be expressed as below:

$$\varphi = \min \frac{\dfrac{1}{m} \sum_{i=1}^{m} \dfrac{\overline{x}}{x_{ik}}}{1 + \dfrac{1}{a+b} \left(\sum_{r=1}^{a} \dfrac{\overline{y^e}}{y_{rk}^e} + \sum_{h=1}^{b} \dfrac{\overline{y^n}}{y_{hk}^n} \right)}$$

$$\text{Subject to} \, \overline{x} \geq \sum_{j=1,\neq k}^{n} x_{ij}\lambda_j, \quad i = 1,\ldots,m$$

$$\overline{y^e} \leq \sum_{j=1,\neq k}^{n} y_{rk}^e \lambda_j, \quad r = 1,\ldots,a$$

$$\overline{y^n} \leq \sum_{j=1,\neq k}^{n} y_{hk}^n \lambda_j, \quad h = 1,\ldots,b$$

$$\lambda_j \geq 0, \quad j = 1,\ldots,n, j \neq 0$$

$$\overline{x} \geq x_k, \, \overline{y^e} \geq y_k^e, \, \overline{y^n} \geq y_k^n$$

(3.3)

3.3.2 Metafrontier production function

Due to certain differences between the factors and the environment in China's coastal areas, the production frontiers corresponding to each area are also different. If a common production frontier is nonetheless used for the efficiency analysis, there will be a certain deviation in the analytical results and the real marine resource utilization efficiency in each area may not be reflected truthfully. To solve this problem, Battese et al. (2004) proposed the metafrontier production function analysis method, which first divides DMUs into different groups according to certain standards, and then builds the metafrontier and group frontier of DMUs according to the DEA method and linear programming, after which the efficiency values of each DMU under the metafrontier technological level and the group frontier technological level are respectively calculated; at the same time, their corresponding metatechnology ratios are also calculated and compared.

The metafrontier technology set that contains undesirable output is defined as T^m:

$$T^m = \left\{ (x, y^e, y^n) : x \ge 0, \ y^n \ge 0; \ x \text{ can produce } (y^e, y^n) \right\} \tag{3.4}$$

The corresponding production possibility set is defined as P^m:

$$P^m = \left\{ (y^e, y^n) : (x, y^e, y^n) \in T^m \right\} \tag{3.5}$$

The metadistance function is:

$$0 \le D^m(x, y^e, y^n) = \sup_\lambda \left\{ \lambda > 0 : (x/\lambda) \in P^m(y^e, y^n) \right\} \le 1 \tag{3.6}$$

This chapter divides China's coastal areas into three groups: the northern coastal region, the eastern coastal region, and the southern coastal region, while the group technology set T is defined as below:

$$T = \left\{ (x_i, y_i^e, y_i^n) : x_i \ge 0, \ y_i^n \ge 0; \ x_i \text{ can produce } (y_i^e, y_i^n) \right\} \tag{3.7}$$

The corresponding group technology efficiency (GTE) is:

$$0 \le D^i(x_i, y_i^e, y_i^n) = \sup_\lambda \left\{ \lambda > 0 : (x_i/\lambda) \in P^i(y_i^e, y_i^n) \right\} \le 1 \tag{3.8}$$

Moreover, the metafrontier technology is the envelopment frontier of the group frontier technology, that is, $T^m = \left\{ T^1, T^2, T^3 \right\}$. As shown in Fig. 3.1, the curve of metafrontier envelopment is higher than that of the group frontiers.

The metatechnology rate (MTR) under the metafrontier reflects the gap between the group frontier technology levels and the metafrontier technology level. The larger the MTR, the smaller the gap between the group frontier technology levels and the metafrontier technology levels; the smaller the MTR, the larger the gap and the lower the actual technological efficiency. This is expressed as below:

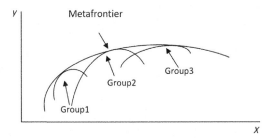

Figure 3.1 Envelopment curves of metafrontier and group frontiers.

$$\mathrm{TE}^i(x_i, y_i^e, y_i^n) = \frac{1}{D^i(x_i, y_i^e, y_i^n)}, \quad i = 1, 2 \tag{3.9}$$

$$0 \le \mathrm{MTR} = \frac{D^m(x, y^e, y^n)}{D^i(x, y_i^e, y_i^n)} = \frac{\mathrm{MTE}}{\mathrm{GTE}} \le 1 \tag{3.10}$$

$$\mathrm{MTE} = \mathrm{MTR} \times \mathrm{GTE} \tag{3.11}$$

According to the research method of Chiu, Liou, Wu, and Fang (2012), the metafrontier inefficiency of each area is decomposed to technological inefficiency (TIE) and management inefficiency (MIE):

$$\mathrm{IE} = 1 - \mathrm{MTE} = \mathrm{TIE} + \mathrm{MIE} \tag{3.12}$$

$$\mathrm{TIE} = \mathrm{GTE} - \mathrm{MTE} \tag{3.13}$$

$$\mathrm{MIE} = 1 - \mathrm{GTE} \tag{3.14}$$

where TIE is the efficiency loss brought about by the gap in production technological levels, whereas MIE is the efficiency loss that is caused by gaps in internal management ability under the same technological level.

3.3.3 Theil index and its decomposition

Based on measuring the utilization efficiency of China's marine resources, the Theil index is used to analyze the source of regional efficiency gaps. The Theil index divides the overall regional efficiency gap into two parts: the intraregional gap and the interregional gap and can analyze the contribution rate of each to the overall gap. The value range of the Theil index is [0,1]. The smaller the Theil index is, the smaller the regional gap is. The greater the Theil index, the greater the regional disparity. The Theil index and its decomposition formula are obtained by referring to Shorrocks (1980), as shown in formulas (3.15)−(3.18):

$$\mathrm{Theil} = \mathrm{Theil}_w + \mathrm{Theil}_b = \frac{1}{11} \sum_{i=1}^{11} \frac{y_i}{\overline{y}} \ln \frac{y_i}{\overline{y}} \tag{3.15}$$

$$\mathrm{Theil}_p = \frac{1}{n_p} \sum_{i=1}^{n_p} \frac{y_{pi}}{\overline{y}_p} \ln \frac{y_{pi}}{\overline{y}_p} \tag{3.16}$$

$$\mathrm{Theil}_w = \sum_{p=1}^{3} \left(\frac{n_p}{11} \frac{\overline{y}_p}{\overline{y}} \right) \mathrm{Theil}_p \tag{3.17}$$

$$\text{Theil}_b = \sum_{p=1}^{3} \frac{n_p}{11} \left(\frac{\bar{y}_p}{\bar{y}}\right) \ln \left(\frac{\bar{y}_p}{\bar{y}}\right) \tag{3.18}$$

where $n_p(P = 1, 2, 3)$ refers to the number of provinces and cities in the eastern, southern, and northern coastal regions, respectively; $\frac{\bar{y}_p}{\bar{y}}$ refers to the ratio of the average marine resource utilization efficiencies of the three areas and the average overall marine resource utilization efficiency; Theil refers to the overall gap in marine resource utilization efficiency; Theil_p refers to the gap of marine resource utilization efficiency for each province or city within each region (the eastern, southern, and northern coastal regions); and Theil_w and Theil_b respectively refer to the intraregional and interregional gaps of coastal areas.

3.4 Estimation of marine resource utilization efficiency

3.4.1 Data selection and analysis

To measure the utilization efficiency of China's marine resources more comprehensively, this chapter selects the marine data of 11 coastal areas in China from 2006 to 2015 for empirical research, including three types of variables: input index, desirable output index, and undesirable output index. The data were mainly drawn from the China Statistical Yearbook and the China Ocean Yearbook (Table 3.1).

This chapter selects the marine resources utilization capital stock and the number of sea-related employees as the capital input index and labor

Table 3.1 Marine resource utilization efficiency evaluation index system.

Type	Primary index	Secondary index
Input index	Capital	Marine resources utilization capital stock (CNY100 million)
	Labor force	Number of sea-related employees (10,000 persons)
Desirable output	Economic benefits	Gross ocean product (GOP; CNY100 million)
Undesirable output	Environmental pollution	Total amount of industrial wastewater that directly discharge into the ocean (10,000 tons)

input index, respectively, for which the calculation of the marine resources utilization capital stock is as below:

$$\text{Marine resources utilization capital stock}$$
$$= (\text{GOP in coastal areas}/\text{GDP in coastal areas})$$
$$\times \text{Capital stock in coastal areas}$$

The estimation of the capital stock in coastal areas takes the method of Shan (2008) for reference with the formula for calculation as below:

$$K_t = K_{t-1}(1 - \delta) + I_t.$$

1. Determine the current year investment data I_t: this is represented by the time series of gross fixed capital formation.
2. Construct the investment price index: calculate the implicit investment deflator to replace the fixed assets investment price index, using the official data of national and provincial gross fixed capital formation and its direct subsidiaries.
3. Determine the depreciation rate δ: according to formula $S = (1 - D)^T$, replace capital efficiency S by the statutory residue value rate and $\delta = 10.96\%$ is obtained.
4. Determine the base period capital stock K: according to $\Delta K / K = \Delta I / I$, the investment price index in coastal areas in 2000 is used to calculate the capital stock in coastal areas, taking the year 2000 as the base period; meanwhile, the GOP in coastal areas and the GDP in coastal areas taking 2000 as the base period are used to calculate the marine resources utilization capital stock.

The desirable output represents the output that brings benefits to the local area through the utilization of marine resources in a certain period. Since the gross ocean product in coastal areas can accurately reflect the desirable output of the ocean, it is included in the analysis. For consistency with the capital stock, it is converted by taking the year 2000 as the base period.

Due to the large proportion of industrial wastewater in environmental pollution in coastal areas, this chapter selects the total amount of industrial wastewater that is discharged directly into the ocean in coastal areas as an proxy index of undesirable output.

Table 3.2 shows the average levels of input and output of marine resource utilization in coastal areas. First, from the perspective of desirable and undesirable outputs, the areas with the top five desirable outputs are

Table 3.2 Average level of input and output in coastal areas.

Area	Y	WD	K	L
Tianjin	2301.4	289.6	6284.7	146.5
Hebei	1051.7	1157.0	2586.2	82.3
Liaoning	2049.1	27357.7	5302.1	278.2
Shanghai	4267.7	9183.4	8794.3	180.9
Jiangsu	2484.4	1018.9	5510.0	165.9
Zhejiang	2608.4	9914.0	5708.9	363.9
Fujian	3309.9	59011.1	7382.7	368.6
Shandong	5476.2	8591.2	13085.1	454.0
Guangdong	6693.4	6742.1	11575.0	717.1
Guangxi	501.5	4157.3	1396.6	97.8
Hainan	466.7	3148.6	1366.8	114.4

Note: Y represents the marine GOP and the unit is CNY100 million; WD stands for the total amount of industrial wastewater discharged directly into the ocean and the unit is 10,000 tons; K represents the marine economy capital stock and the unit is CNY100 million; and L stands for the number of sea-related employees and the unit is 10,000 persons.

Hebei, Zhejiang, Liaoning, Jiangsu, and Shandong, whereas the areas with the top five undesirable outputs are Fujian, Liaoning, Zhejiang, Shanghai, and Shandong. Three areas out of the top five areas with desirable outputs are also included in the top five areas with undesirable output. The change in sequence indicates the differences in the abilities of these different areas in terms of environmental governance. However, we can still conclude that the expansion of the marine industry is, to some extent, accompanied by a rise in environmental governance pressure. From the perspective of input and gross regional product (GRP), there are great differences between the ranking of inputs and the ranking by GRP in each area. A certain gap can be seen in the technical level of the marine economy in different areas, and there are differences in the technical environment.

Table 3.3 shows the proportion of annual average input and output of each area to the total annual average input and output of all areas. As there are similarities inside each economic region and certain differences among different regions, this chapter divides the 11 coastal areas into the northern coastal region (including Tianjin, Shandong, Hebei, and Liaoning), the eastern coastal region (including Shanghai, Jiangsu, and Zhejiang) and the southern coastal region (including Fujian, Guangdong, Guangxi, and Hainan). The northern coastal region and the southern coastal region have the highest proportions of input and desirable outputs, followed by the eastern coastal region. However, the southern coastal

Table 3.3 Proportion of annual average input and output of each area in the total annual average input and output of all areas (%).

Area	K	L	Y	WD
Tianjin	9.11	4.93	7.37	0.22
Liaoning	7.69	9.37	6.57	20.95
Hebei	3.75	2.77	3.37	0.89
Shandong	18.97	15.29	17.55	6.58
Jiangsu	7.99	5.59	7.96	0.78
Zhejiang	8.27	12.25	8.36	7.59
Shanghai	12.75	6.09	13.67	7.03
Guangdong	16.78	24.15	21.45	5.16
Fujian	10.70	12.41	10.61	45.19
Hainan	1.98	3.85	1.50	2.41
Guangxi	2.02	3.29	1.61	3.18
Northern coastal region	39.51	32.36	34.85	28.64
Eastern coastal region	29.01	23.93	29.99	15.41
Southern coastal region	31.48	43.71	35.15	55.95

region has the highest proportion of undesirable output (55.95%), whereas the eastern coastal region has the lowest (15.41%), indicating that the environmental governance capacity in the southern coastal region is relatively poor, whereas that in the northern and eastern coastal regions is strong.

3.4.2 Calculation and analysis of marine resource utilization efficiency

Empirical analysis is carried out using the superefficiency SBM model and metafrontier function. Based on the different frontiers mentioned above, we can use MaxDEA software to calculate the marine resource utilization efficiency of each area, on the premise of desirable and undesirable outputs being equal, based on CRS. In the estimation process, SBM can be divided into three basic types, namely, CRS, VRS, and general returns to scale (GRS). Due to the condition of DMU linear programming being infeasible in the estimation process of VRS and GRS, the CRS is selected in this chapter's efficiency calculations. Statistical descriptions of input and output indexes of marine resources utilization from 2006 to 2015 are listed in Table 3.4.

Table 3.5 shows that the average group technology efficiencies of the northern, eastern, and southern coastal regions from 2006 to 2015 are, respectively, 1.029, 1.118, and 0.762, indicating that under their

Table 3.4 Statistical descriptions of marine input and output indexes from 2006 to 2015.

Variable	Observed values	Mean value	Standard error	Maximum value	Minimum value
Marine economy capital stock (CNY100 million)	110	6272.0	4699.9	23497.6	476.6
Number of sea-related employees (10,000 persons)	110	270.0	207.3	860.3	81.5
GOP (CNY100 million)	110	2837.3	2174.5	10486.5	268.8
Total discharge of industrial wastewater in coastal areas (10,000 tons)	110	11870.1	17973.6	107994.4	0.5

respective technological levels, the gap in efficiency between the northern and eastern coastal regions is smaller, and both are fully effective; meanwhile, the efficiency value of the eastern coastal region ranks higher than that of the northern coastal region, whereas the efficiency in the southern coastal region greatly lags behind them. In addition, the metafrontier technology efficiencies (marine resources utilization) of the three regions show a similar situation, that is, the average metaefficiency gaps of the northern and eastern coastal regions are small, respectively, 0.811 and 0.895, whereas the efficiency in the southern coastal region, which is 0.582, lags greatly behind them. It can be seen that regardless of the metafrontier technology efficiency or the group frontier technology efficiency, resources utilization in the southern coastal region lags far behind the northern and eastern coastal regions, with resource utilization in the eastern coastal region being higher than that in the northern coastal region, meaning that both resource utilization and environmental protection in the eastern coastal region are more effective, as compared with the northern and southern regions.

By using formulas (3.15)−(3.18) for calculation, we can obtain the results in Table 3.6.

As can be seen from Table 3.6, the utilization efficiency of marine resources in China increased from 0.1058 in 2006 to 0.1635 in 2015 with an average annual increase rate of 4.45%. Meanwhile, in general, the interregional gap and the intraregional gap both increased, indicating that the gaps for all areas were expanding as a whole. This finding means that

...metatechnological rate, and the change rate and inefficiency of each area between 2006 and 2015.

Northern coastal region	MTE		GTE		MTR		MTR-R	IE	TIE and its proportion		MIE and its proportion	
	AVE	SD	AVE	SD	AVE	SD	AVE	AVE	AVE	AVE	AVE	AVE
Tianjin	1.516	0.431	1.709	0.314	0.877	0.127	−0.042	0	0.193	100.00%	0	0.00%
Hebei	0.73	0.244	0.969	0.163	0.741	0.14	−0.016	0.293	0.239	74.43%	0.082	25.57%
Liaoning	0.4	0.022	0.492	0.024	0.814	0.062	−0.022	0.6	0.093	15.42%	0.508	84.58%
Shandong	0.6	0.054	0.946	0.17	0.663	0.184	−0.047	0.4	0.346	82.26%	0.075	17.74%
AVE	0.811		1.029		0.774		−0.032	0.323	0.218	56.72%	0.166	43.28%
SD	0.489		0.504		0.092		0.015	0.25	0.105	0.367	0.231	0.367

Eastern coastal region	MTE		GTE		MTR		MTR-R	IE	TIE and proportion		MIE and its proportion	
	AVE	SD	AVE	SD	AVE	SD	AVE	AVE	AVE	AVE	AVE	AVE
Shanghai	1.19	0.089	1.275	0.165	0.94	0.059	0.02	0	0.085	100.00%	0	0.00%
Jiangsu	1.01	0.15	1.583	0.384	0.667	0.138	0.031	0.04	0.573	93.45%	0.04	6.55%
Zhejiang	0.486	0.037	0.495	0.042	0.983	0.023	0.008	0.514	0.009	1.71%	0.505	98.29%
AVE	0.895		1.118		0.863		0.02	0.185	0.222	65.05%	0.182	34.95%
SD	0.366		0.561		0.171		0.012	0.286	0.306	0.55	0.281	0.55

Southern coastal region	MTE		GTE		MTR		MTR-R	IE	TIE and proportion		MIE and its proportion	
	AVE	SD	AVE	SD	AVE	SD	AVE	AVE	AVE	AVE	AVE	AVE
Fujian	0.477	0.017	0.668	0.186	0.749	0.143	−0.022	0.523	0.191	36.26%	0.336	63.74%
Guangdong	1.123	0.035	1.574	0.202	0.726	0.108	−0.029	0	0.451	100.00%	0	0.00%
Guangxi	0.403	0.055	0.449	0.062	0.899	0.049	−0.006	0.597	0.046	7.70%	0.551	92.30%
Hainan	0.326	0.023	0.359	0.02	0.91	0.054	0.009	0.674	0.032	4.81%	0.641	95.19%
AVE	0.582		0.762		0.821		−0.012	0.449	0.18	37.19%	0.382	62.81%
SD	0.366		0.556		0.097		0.017	0.305	0.194	0.442	0.285	0.442

AVE, average value; *GTE*, group frontier technology efficiency; *IE*, inefficiency; *MIE*, management inefficiency; *MTE*, metatechnological efficiency; *MTR*, metatechnology efficiency; *MTR-R*, change rate of metatechnological rate; *SD*, standard error; *TIE*, technological gap inefficiency.

Table 3.6 Regional differences of marine resource utilization efficiency and decomposition.

Year	Overall gap	Interregional gap Value	Contribution rate (%)	Intraregional gap Value	Contribution rate (%)	Eastern Contribution rate (%)	Southern Contribution rate (%)	Northern Contribution rate (%)
2006	0.1058	0.0141	13.31	0.0917	86.69	24.15	39.58	22.96
2007	0.0998	0.0141	14.15	0.0857	85.85	23.30	38.74	23.80
2008	0.1106	0.0186	16.82	0.0920	83.18	23.61	42.28	17.28
2009	0.1188	0.0065	5.51	0.1122	94.49	28.55	41.09	24.85
2010	0.1181	0.0170	14.38	0.1011	85.62	21.13	40.34	24.15
2011	0.1583	0.0182	11.48	0.1401	88.52	9.84	29.86	48.82
2012	0.1541	0.0199	12.89	0.1342	87.11	11.44	29.38	46.28
2013	0.1527	0.0212	13.89	0.1315	86.11	10.87	28.57	46.66
2014	0.1559	0.0198	12.69	0.1361	87.31	11.58	28.95	46.78
2015	0.1635	0.0210	12.87	0.1424	87.13	11.41	27.53	48.19
Average	0.1338	0.0170	12.80	0.1167	87.20	17.59	34.63	34.98

the efficiency gap in the utilization of marine resources within each region was widening, which was on the contrary to China's goal of coordinated regional development. As for the source of these regional gaps, the overall regional gap in the utilization efficiency of marine resources mainly came from the intraregional gap, the contribution rate of which remained at 83.00%–94.50%. Meanwhile, the contribution of the interregional gap will always be below that of the intraregional gap in the short term. From the perspective of intraregional gaps, the gap between areas in the southern and the northern regions was relatively large, 34.63% and 34.98%, respectively, followed by 17.59% in the eastern region. Judging from the decomposition results, the improvement of utilization efficiency of marine resources in China should focus on narrowing the gaps among areas within the eastern, southern, and northern coastal regions. However, it is worth noting that the gap among areas within the eastern coastal region decreased during the sample period, indicating that coordinated development was emphasized in the eastern region. The gap among areas within the northern region rose gradually from 2008 and soared in 2011, which was mainly due to the significant decline of marine resources utilization efficiencies in Shandong and Hebei between 2008 and 2011.

For further analysis, mean values of metatechnology efficiency and group frontier technology efficiency of 11 coastal areas from 2006 to 2015 are sorted according to the calculation results, as shown in Fig. 3.2.

From Fig. 3.2, we can see directly that the marine resources utilization efficiencies of four out of the 11 observed areas exceeded one, these areas being Tianjin, Shanghai, Guangdong, and Jiangsu, which means that the inputs and outputs in these four areas had reached the optimal level and the protection of resources and environment there was effectively maintained while developing the marine economy. Efficiencies in the other seven areas were all at low levels, and inputs and outputs should be changed to improve their efficiencies effectively. Results under the group frontier were same as those under the metafrontier. The marine resources utilization efficiencies in Tianjin, Jiangsu, Guangdong, and Shanghai all exceeded one, among which Tianjin belongs to the northern coastal region, Jiangsu, and Shanghai to the eastern region and Guangdong to the southern region. This shows that at present, China's eastern coastal region pays more attention in the development of the marine economy to making full use of resources and environmental protection, and the marine economic development mode there is relatively sustainable; in the northern and southern regions, the marine economic development modes are

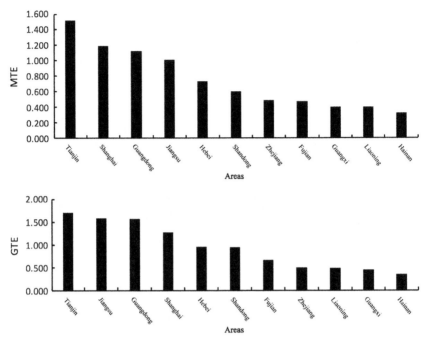

Figure 3.2 Metatechnology efficiencies and group frontier technology efficiencies of 11 coastal areas.

comparatively underdeveloped, hence, they should develop in a coordinated way and take into account both economic benefits and resource and environmental benefits.

According to the changes in time series of marine resource utilization efficiency in Fig. 3.3:

1. From 2006 to 2015, the utilization efficiency of marine resources in the eastern coastal region was generally higher than that in the northern and southern regions, under both the metafrontier technology and group frontier technology.

2. Under the metafrontier technology level, the marine resource utilization efficiency in the eastern coastal region decreased slightly in general and fluctuated greatly between 2008 and 2010, which may be because of the impacts of the financial crisis. In addition, there is also research (Wang, 2014) showing that the inputs and outputs of the marine economy had significant impacts on productivity, and the influence degree of the output was more significant as compared with that of the input. Therefore we can judge that the decrease of marine

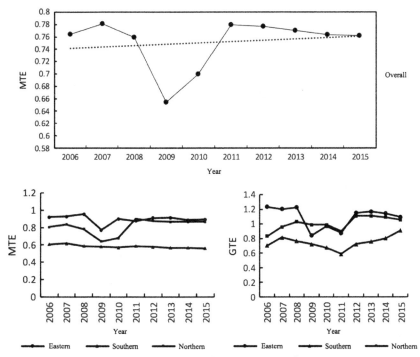

Figure 3.3 Variation trends of averages of MTE and GTE of China's marine economy between 2006 and 2015.

output in recent years exacerbated the decline of productivity. The marine resource utilization efficiency in the northern coastal region between 2006 and 2015 reached its peak value of 0.899 in 2011 and generally showed an upward trend; there was a notable decline in 2009, which may because the 11th 5-year plan drew to a close and the governmental governance on marine resources utilization was loosened. There is also research (Wang, 2014) showing that technological degradation was directly related to the decline of marine resource utilization efficiency. Changes in the southern coastal region were not significant and the overall trend was almost horizontal with the highest value reaching 0.616 in 2007. It can be seen from Fig. 3.3 that the efficiency values of the three regions were all less than one, which means that there was still extensive room for improvement in marine resources utilization in China; especially for the northern and southern coastal regions, there was still much room for resource conservation and efficiency improvement.

3. Under the group frontier technological level, variation trends in the efficiencies in the three regions were similar to those under the meta-frontier. Moreover, taking an efficiency value of one as the dividing line, the eastern coastal region was above the line all the time, except in 2009, 2010, and 2011, indicating that the inputs and outputs of marine resources utilization in the eastern region during the sample period were both completely effective, except for the decline in 2008 because of the economic crisis, with a gradual restoration in 2009. Meanwhile, the efficiency values of the northern region were above the dividing line in the late sample period. However, the efficiency values of the southern region were always below the dividing line during the sample period, indicating that the overall efficiency of marine resources utilization in this region was low and that there is much room for improvement.

4. Overall, the utilization efficiency of China's marine resources fluctuated greatly before 2011, declined before 2009, and increased from 2009 to 2011. As can be seen from Fig. 3.4, the overall utilization efficiency of China's marine resources still showed a slight increase from 2006 to 2015.

As seen in Fig. 3.4, the averages of MTRs in the northern, eastern, and southern regions from 2006 to 2015 were 0.774, 0.863, and 0.821, respectively, indicating that the technological levels in the eastern and southern coastal regions were up to 86.3% and 82.1% of the metafrontier level, respectively, while the level in the northern coastal region only

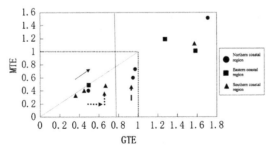

Figure 3.4 GTE–MTE matrix of the northern, eastern, and southern coastal regions. *Note:* ⟶ refers to the improvement direction of efficiencies of the eastern coastal area; ⋯⋯▶ refers to the improvement direction of efficiencies of the northern and southern coastal regions to the left of the dividing line; – ⟶ refers to the improvement direction of efficiencies of the northern coastal region to the right of the dividing line.

reached 77.4%. According to group frontier technical analysis, the northern and eastern coastal regions were in a fully effective status and the room for improvement for the southern coastal region was 23.8%. Under the metafrontier technology, the room for efficiency improvement was, in increasing order, the eastern coastal region < the northern coastal region < the southern coastal region (10.5% < 18.9% < 41.8%), showing that the technological level in the southern coastal region clearly lagged behind the levels in the eastern and northern coastal regions.

By making a 2D matrix of group efficiencies and metaefficiencies for the northern, eastern, and southern coastal regions, we can see that in terms of the scope of the efficiency value being <1, the eastern region was above the diagonal, whereas the northern and the southern regions were below the diagonal, indicating that the technological level in the eastern region was basically the same as the metafrontier technological level, whereas the technological levels in the northern and southern regions lagged behind that of the eastern region. Then, by taking the average value of the metaefficiency as the dividing line (vertical solid line in Fig. 3.4) for analysis, we can see that the key for improving the efficiency in the northern coastal region, on the right of the dividing line, is to improve its technological level; the key for improving the efficiency in the eastern coastal region, on the left of the dividing line, is to improve its management level; while both technological levels and management levels in the northern and southern coastal regions are in need of improvement.

Moreover, the variation conditions of the gaps between regions and the optimal production technology level are analyzed through calculation of the variation in the MTR-R. It can be seen from Table 3.5 that the average value of MTR-R in the eastern coastal region was positive and the standard deviation was low, indicating that the technical gap between the eastern coastal region and the metafrontier was gradually shrinking, and there are few internal differences in this process. However, the MTR-Rs of the northern and southern coastal regions were negative, indicating that the gaps between the northern and southern coastal regions and the metafrontier had been expanding, mainly because of technological degradation in certain individual areas. Judging from each area, the MTR-R of Shanghai, Jiangsu, and Zhejiang in the eastern coastal region from 2006 to 2015 had annual average increases of 2%, 3%, and 0.8%, respectively. In the northern coastal region group, degradation in Tianjin and Shandong were more serious than in other areas; in the southern region group, there was technological degradation in Guangdong and Guangxi, while technological progress was seen in Hainan.

The findings of Wang (2014) showed that the lowering of technology levels in coastal areas was to some extent derived from the reduction of marine capital input. In fact, there has been an inevitable phenomenon of excessive capital investment in China. Since the start of the 21st century with the gradual development of the marine economy, the marine input and output in each area was reduced, along with the regulation of marine capital input. Under the condition that each DMU has formed its best practice over a long time and cannot easily be extended further, the decline of frontier technological levels follows. The changed of technological level in each area may have resulted from different levels of regional economic development. Underdeveloped areas have certain late-mover advantages, meaning that technological levels can be greatly promoted by policy incentives, while the room for improvement for developed areas is small, due to their high original technological levels, and their developmental pace was therefore slower.

To further explore the internal causes of low efficiency in marine resources utilization and analyze the sources of gaps between the actual and the potential optimal technological levels in different regions, we divide the efficiency losses in the three regions into two dimensions: technological gap inefficiency and MIE. The results are shown in Table 3.5. The total efficiency loss in the eastern coastal region was 0.185 and its technological gap inefficiency value reached a maximum of 0.222 of the three groups with a contribution rate of 65.05%; its MIE value was 0.182 with a contribution rate of 34.95%. Therefore the future development of the eastern coastal region should focus on improving the technological level first and then the management level. The situation in the southern region was contrary to that that of the eastern region. The total value of efficiency loss was 0.449, the highest of the three groups. Its MIE value also attained the maximum value of 0.382 of the three groups with a contribution rate of 62.81%. However, its TIE gap value was the smallest, only 0.18, with a contribution rate of 37.19%. Therefore the low efficiency of the southern coastal region comes mainly from its low management level. Thus in future development, the improvement of management ability should be the main consideration, while the improvement of the technological level should not be ignored. The efficiency loss in the northern coastal region was 0.323; its TIE was 0.218 with a contribution rate of 56.72% and its MIE value was 0.166, with a contribution rate of 43.28%. Hence, for the northern coastal region, improving technology levels and management efficiency in tandem should be considered in the future.

3.5 Dynamic evolution of marine resource utilization efficiency

3.5.1 Modeling

3.5.1.1 Kernel density estimation method

The density function of random variables can be estimated using the kernel density estimation approach. Assuming that random variables $X_1, X_2, \ldots, X_N$ are uniformly distributed and that the density function is $f(x)$, the empirical distribution function is as follows:

$$F_N(y) = \frac{1}{N} \sum_{i=1}^{N} I(X_i \le y) \tag{3.19}$$

where N refers to the sum of the observed values; $I(z)$ refers to the index function; and z refers to the corresponding condition relational expression. When z is true, $I(z) = 1$; otherwise, $I(z) = 0$.

Set the kernel function:

$$\text{The } \mathrm{il}\eta_o(x) = \begin{cases} 1/2 & -1 \le x \le 1 \\ 0 & \text{otherwise} \end{cases} \tag{3.20}$$

Then, the corresponding kernel density estimation is as below:

$$f(x) = \frac{1}{hN} \sum_{i=1}^{N} \eta\left(\frac{x - X_i}{h}\right) \tag{3.21}$$

where h refers to bandwidth and η refers to the kernel function. The Epanechnikov kernel function is selected:

$$\eta(u) = \frac{p(p+2)}{2 S_p} \left(1 - u_1^2 - u_2^2 - \cdots - u_p^2\right) \tag{3.22}$$

where $S_p = 2\pi^{r/2}/\Gamma(p/2)$; when $p = 1$, $\Gamma(u) = 0.75(1 - u^2)I(|u| \le 1)$.

3.5.1.2 Markov chain method

Markov chain estimation is used to evaluate the dynamic evolution of marine resource utilization efficiency, based on its fluidity in different areas. Markov chain estimation assumes that the sequence of variables has the feature of "nonaftereffect," that is, when its present state is known, the distribution of its future state has nothing to do with its past state. Assume that there is a random process $x(t)$; if the probability of transforming from the state i at t to the state j at $t+1$ is p_{ij} and the probability of being in the state s_i at t is $a_i(t)$, then the evolution is captured by:

$$a_i(t+1) = \sum_{i=1}^{n} a_i(t)p_{ij}(i = 1, 2, \ldots, n) \tag{3.23}$$

This indicates that the state probability of each area at $t+1$ is only related to the state probability of each area at t and its transition probability, and is not related to its state before t. In this chapter, the probability distribution of marine resource utilization efficiency of each coastal area in the year t is expressed as a $1 \times K$ state probability vector P_t that written as $P_t = (P_1(t), P_2(t), P_3(t), \ldots, P_n(t))$. Table 3.7 presents the transition probability matrix $(K \times K)$ of marine resource utilization efficiency for each area.

In Table 3.7, p_{ij} represents the probability of a type-i coastal area transforming to a type-j area during the whole research period, for which the computing formula is $p_{ij} = \frac{n_{ij}}{n_i}$; n_{ij} represents the sum of the quantity of all areas that changed from type-i to type-j during the whole research period; and n_i refers to the sum of quantity of type-i areas during the whole research period. If one type-i coastal area remains in type i the next year, then the transfer of this area is stable; if one coastal area transforms to a type that is higher than the previous year, it means that this area has experienced an upward transfer; otherwise, the marine resource utilization efficiency in this area is in downward transfer.

Table 3.7 Markov transition probability matrix.

Type	Low level	Medium and low level	Medium level	High and medium level	High level
Low level	P_{11}	P_{12}	P_{13}	P_{14}	P_{15}
Medium and low level	P_{21}	P_{22}	P_{23}	P_{24}	P_{25}
Medium level	P_{31}	P_{32}	P_{33}	P_{34}	P_{35}
High and medium level	P_{41}	P_{42}	P_{43}	P_{44}	P_{45}
High level	P_{51}	P_{52}	P_{53}	P_{54}	P_{55}

Under the condition that the state transition probability matrix P possesses certain stability properties as time changes, then the marine resource utilization efficiency distribution after n stages $a(t + n)$ can be expressed as $a(t + n) = p^n a(t)$. When $n \to \infty$, if $a(t + n)$ shows convergence, the stationary distribution of marine resource utilization efficiency of each area can be obtained, that is, $\pi = (\pi_1, \pi_2, \ldots, \pi_n)$. If π shows a state of concentrated distribution, it will be considered that the gap of marine resource utilization efficiency for each coastal area can be eliminated slowly; if π is in a dispersed state or there is no solution at all, it will be assumed that the gap of marine resource utilization efficiency for each coastal area cannot be eliminated or even represents a state of polarization. Therefore through the analysis of these stationary distribution states, the development trend of marine resource utilization efficiency of each coastal area can be predicted.

3.5.2 Results analysis
3.5.2.1 Kernel density estimation results
By calculating the kernel density of the marine resource utilization efficiency of coastal areas during the research period, the distribution state can be judged. If the kernel density distribution diagram presents as a single-peak state, this shows that the marine resource utilization efficiency of coastal areas converges to a certain point. If the kernel density distribution shows a multipeak state, then it shows that the marine resource utilization efficiency of coastal areas is in a state of multiple convergence, that is, there is multipolar differentiation of marine resource utilization efficiencies of coastal areas. In addition, the peak value also reflects the distribution of the marine resource utilization efficiency. If the peak value decreases with time, the gap of marine resource utilization efficiency in every area increases with time, and the relative concentration ratio decreases as well. On the contrary, if the peak value increases with time, this indicates that the gap of marine resource utilization efficiency for each area decreases with time and the relative concentration ratio increases. The specific change forms of distribution of density function and their corresponding meanings are shown in Table 3.8 and Fig. 3.5.

The calculation results of marine resource utilization efficiency of each region are input into Eviews7.2 software, and the commonly used Epanechnikov kernel function is selected to calculate the kernel density distribution of marine resource utilization efficiency in each year. In order to observe more clearly the variation trend in the density distribution of

Table 3.8 Change forms of density function distribution and corresponding meanings.

	Gap increases	Gap decreases
Peak height	Shorten	Heighten
Peak width	Widen	Narrow down
Peak skewness	Left skewed	Right skewed
Peak quantity	Increase	Decrease

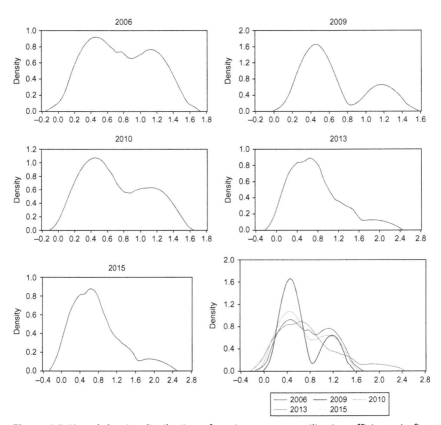

Figure 3.5 Kernel density distribution of marine resource utilization efficiency in five specific years.

marine resource utilization efficiency, the kernel density distribution diagram of the marine resource utilization efficiency in 2006, 2009, 2010, 2013, and 2015 is drawn, in order to reflect the interannual variation of the marine resource utilization efficiency, as shown in the following Fig. 3.6.

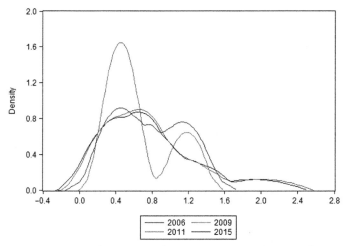

Figure 3.6 Kernel density distribution of marine resource utilization efficiency.

First, from the perspective of location, the kernel density distribution curves of the five individual years as a whole shows a trend moving to the right, indicating that the marine resource utilization efficiency of China presents an increasing trend. Second, the peak value tends to move to the right, while the trailing part on the right side becomes increasingly large over time, indicating that the marine resource utilization efficiency increases constantly, but the height of the peak gradually decreases, indicating that the degree of concentration of efficiency of every area gradually decreases, while the dispersion gradually increases. Finally, from the perspective of shape, from 2006 to 2015 the marine resource utilization efficiency in China basically presents a two-peak distribution, indicating that polarization of marine resource utilization efficiency exists throughout the research period. As can be seen from the Fig. 3.6, there is a transfer trend from a double-peak to a single-peak shape during 2011 and 2015, which indicates that this polarization of efficiency was weakening. As the years advanced from 2006 to 2009, the right tail tended to move to the left, indicating that the overall efficiency declined, while from 2009 to 2011, the right tail tended to move to the right, indicating that the overall utilization efficiency showed an upward trend. Changes in other years were small. Changes of the growth rates of marine resource utilization efficiency in low-efficiency areas were relatively stable.

3.5.2.2 Markov chain analysis results

To further analyze the long-term development trend of marine resource utilization efficiency in China's coastal areas, this chapter adopts the Markov chain estimation method.

Using the quartile method, the marine resources utilization efficiencies of coastal areas in the range [0,1] are divided into four types, and areas with marine resource utilization efficiency >1 are classified into the fifth type. Please see Table 3.9 for details.

According to the above research findings, the whole research period is divided into three periods, 2006−09, 2009−11, and 2011−15; the Markov transfer matrixes of marine resource utilization efficiency of coastal areas in these three periods are then calculated as shown in Table 3.10.

From Table 3.10, we can see the following:

1. In terms of liquidity. First, the values on the diagonals of the Markov transfer matrixes during the three periods, as well as the whole research period, are higher than values that do not lie on the diagonals most of the time, indicating that the fluidity of marine resource utilization efficiency is low for each of the three periods, as well as over the whole research period; and the probability of transfer of marine resource utilization efficiency to other types is small. Second, between 2006 and 2009, the minimum value on the diagonal of Markov chain transfer matrix is 0 and the maximum value is 1, indicating that the probability of marine resource utilization efficiency of coastal areas in 2009 remaining unchanged is 100% at the highest and 0% at the lowest; between 2009 and 2011, the minimum value on the diagonal is 0.25 and the maximum value is 1, indicating that the probability of marine resource utilization efficiency of coastal areas remaining unchanged in 2011 is 100% at the highest and 25% at the lowest; between 2011 and 2015, the minimum value on the diagonal of Markov chain transfer matrix is 0.625 and the maximum value is 1, indicating that the probability of this efficiency remaining unchanged

Table 3.9 Type and range value.

Type Period	Low level	Medium and low level	Medium level	High and medium level	High level
2006−09	<0.43	0.43−0.48	0.48−0.68	0.68−1.00	>1.00
2009−11	<0.40	0.40−0.47	0.47−0.55	0.55−1.00	>1.00
2011−15	<0.38	0.38−0.48	0.48−0.55	0.55−1.00	>1.00
2006−15	<0.40	0.40−0.48	0.48−0.54	0.54−1.00	>1.00

Table 3.10 Marine resource utilization efficiency and Markov transfer matrix.

Period	Type	Low level	Medium and low level	Medium level	High and medium level	High level
2006−09	Low level	1.000	0.000	0.000	0.000	0.000
	Medium and low level	0.000	1.000	0.000	0.000	0.000
	Medium level	0.167	0.167	0.500	0.167	0.000
	High and medium level	0.000	0.000	1.000	0.000	0.000
	High level	0.000	0.000	0.000	0.133	0.867
	Stationary distribution	0.212	0.121	0.182	0.030	0.455
2009−11	Low level	0.833	0.167	0.000	0.000	0.000
	Medium and low level	0.000	0.667	0.333	0.000	0.000
	Medium level	0.000	0.250	0.250	0.500	0.000
	High and medium level	0.000	0.000	0.000	0.500	0.500
	High level	0.000	0.000	0.000	0.000	1.000
	Stationary distribution	0.273	0.136	0.182	0.091	0.318
2011−15	Low level	1.000	0.000	0.000	0.000	0.000
	Medium and low level	0.250	0.625	0.125	0.000	0.000
	Medium level	0.000	0.000	1.000	0.000	0.000
	High and medium level	0.000	0.000	0.000	1.000	0.000
	High level	0.000	0.000	0.000	0.000	1.000
	Stationary distribution	0.159	0.182	0.114	0.182	0.364
2006−15	Low level	0.947	0.053	0.000	0.000	0.000
	Medium and low level	0.176	0.706	0.118	0.000	0.000
	Medium level	0.000	0.182	0.636	0.182	0.000
	High and medium level	0.000	0.000	0.071	0.857	0.071
	High level	0.000	0.000	0.026	0.026	0.947
	Stationary distribution	0.192	0.172	0.111	0.141	0.384

in 2015 is 100% at the highest and 62.5% at the lowest; during the whole research period, the minimum value on the diagonal of Markov chain transfer matrix is 0.636 and the maximum value is

0.947, indicating that the probability of this efficiency remaining unchanged in 2015 is 94.7% at the highest and 63.6% at the lowest. This means that the state stability is high during all three periods and over the whole research period, especially during the period 2011−15.

2. In terms of span of changes. Changes in the marine resource utilization efficiency are small. Judging from the three subperiods, changes only span from the high level to the high and medium levels; judging from the whole research period, the probability of transfer from the medium level to the high level is 0%, the probability of transfer from the high level to the medium level is 2.6%, and there is no transfer from the low level to the high level. Thus it can be said that the span of changes of marine resource utilization efficiency is small.

3. In terms of distribution. The marine resource utilization efficiency showed obvious convergence over the whole research period and each of the three subperiods. Between 2001 and 2011, the probability of areas that were initially at a low level remaining unchanged in following years is 94.7%, and only 5.3% of areas move upward. In line with the analytical results above, showing that the marine resource utilization efficiency of China's coastal areas presented a small increasing trend during the whole research period, the probability of areas that were at low levels, or at medium or low levels in 2006 moving upward with time are, respectively, 5.3% and 11.8%; there is no transfer from high-level areas and the probability of remaining unchanged is 94.7%. Meantime, according to the above analysis, between 2006 and 2009, the efficiency showed a decreasing trend, which was also observed in the Markov transfer matrix; the probability of high-level areas moving downward was 13.3%, while that of medium- or high-level areas moving downward reached 100%. Moreover, between 2009 and 2011, the probabilities of low-level, medium- or low-level, and medium-level areas in 2009 moving upward with the time lapse were, respectively, 16.7%, 33.3%, and 50%, while the probability of high-level areas remaining unchanged was 100%. This indicates that the marine resource utilization efficiency during this period was in a sharp growth trend.

4. In terms of stationary distribution. It can be seen from the table that over the whole research period, as well as in each of the three subperiods, the marine resource utilization efficiency of every region is still dispersed in five states under the long-term equilibrium condition.

Between 2006 and 2015, the proportion of the stationary distribution of areas with either low or high marine resource utilization efficiencies under the long-term equilibrium state is 57.6%, among which the share of areas with high-level efficiencies was the largest at 38.4%. Therefore if the marine resource utilization efficiency in China continues to develop without change, the gap of marine resource utilization efficiency in every area will still be large and will be hard to improve in a short time.

3.6 Analysis of factors influencing marine resource utilization efficiency

Previously, we combined the superefficiency SBM model with the metafrontier function to calculate the marine resource utilization efficiencies of China's 11 coastal areas and three major regions from 2006 to 2015, and analyzed their development trends and characteristics. To further explore the motivations for China's marine resource utilization efficiency, the symbolic regression model will be used in this chapter to analyze the influencing factors of marine resource utilization efficiency under the metafrontier condition.

3.6.1 Symbolic regression method

Symbolic regression is an evolutionary function discovery method based on genetic programming that was first proposed by Koza (1992). By contrast with traditional regression methods, symbolic regression can determine the parameters and structure of the regression model simultaneously. In symbolic regression, the task is automatically to find the appropriate function form in complex data, whether linear or nonlinear, while determining the coefficient of the function. Compared with other methods, the advantage of symbolic regression is that it can select the best scheme among all the preselected schemes.

3.6.2 Index selection and data source

According to the previous analysis, there are differences in the marine resource utilization efficiencies among China's coastal areas. In this section, based on the analytical frameworks of existing studies and the real environment of marine resource utilization in China's coastal areas, and considering the availability and quantification of data, the following indexes are selected to explore the influencing factors of marine resource utilization efficiency:

3.6.2.1 Marine environment governance (X1)
This index is expressed by the number of regional projects (completed projects in current year) on wastewater and solid waste treatment.

3.6.2.2 Marine technological progress (X2)
Employees engaged in scientific research, number of scientific research institutes and number of research subjects of scientific research institutes are selected as indexes to measure marine technological progress. Considering that the influences of these indexes on marine technological progress are balanced, an average weighting is carried out on these indexes.

3.6.2.3 Sea-related employees' scientific attainment (X3)
The proportion of sea-related professionals of the total sea-related employees is taken to represent the scientific attainment of sea-related employees.

3.6.2.4 Marine resource endowment (X4)
Due to limitations of data availability and quantification, per capita mariculture area (unit: hectare/person) is selected as an index to measure marine resource endowment.

To ensure the accuracy and scientific reliability of the results, normalization processing using the threshold value method is carried out on the selected data. The data are obtained from the China Marine Statistical Yearbook, China Statistical Yearbook, and Marine Statistical Yearbooks of relevant coastal provinces and cities.

3.6.3 Empirical analysis

In this section, the following symbols that appear most frequently in regression models are selected: constant, input variable, addition, subtraction, and multiplication. Eureqa software is then used for the symbolic regression. Symbolic regression does not need to assume a model form and can automatically generate a set of suitable models to build the Pareto frontier. The number of models that can be shown on the Pareto frontier is limited, thus further research should be conducted on these Pareto optimal solutions to find out the factors that appear most frequently in the best models.

The symbolic regression models described in Table 3.11 appear at different Pareto points on the Pareto frontier (18 squares for 18 models) in Fig. 3.7.

In Fig. 3.7, the unfeasible region is represented below the line, the feasible region is above the line, and the Pareto frontier and Pareto points are on the line. For the symbolic regression problem, the higher the complexity, the smaller the model error and the better the model.

According to the factor statistics in the Pareto optimal model, the appearance of each factor is shown in Fig. 3.8, which shows how many models contain each factor and the number of times each factor appears in the model, since a factor may appear more than once in a given model. As can be seen from Fig. 3.8, the order of influences of various factors on the marine resource utilization efficiency is: sea-related employees' scientific attainment > marine technological progress > marine environment governance > marine resource endowment. The sea-related employees' scientific attainment is the most common factor that appears in the Pareto optimal model, while the marine resource endowment appears the least, which means that the sea-related employees' scientific attainment has the strongest influence on marine resource utilization efficiency, while the marine resource endowment has the weakest influence.

The scientific attainment of sea-related employees reflects the level of technological activeness of an industry to a certain extent, and plays a great role in promoting the improvement of marine resource utilization efficiency. This suggests that the improvement of marine resource utilization efficiency comes not only from the research achievements of research institutions, but also the elevation of the technological activity level. Therefore we should be committed to upgrading the scientific attainment of sea-related employees on the premise of considering the real environment of marine resource utilization in every area to improve marine resource utilization efficiency.

The progress of marine technology plays a great role in promoting the development and improvement of marine resource utilization and reflects the scientific and technological level of a region and its ability to transform and apply scientific research results. The improvement of marine resource utilization efficiency cannot be achieved without technological progress in this field.

Table 3.11 Results of symbolic regression.

ID	C	MAE	Models
1	1	0.1591	$y = x3$
2	3	0.1181	$y = 0.648x3$
3	5	0.1165	$y = 0.0277 + 0.606x3$
4	7	0.1129	$y = 0.613x3 + 0.107x1$
5	9	0.1089	$y = 0.0552 + 0.639x3 - 0.186x4$
6	11	0.1050	$y = 0.617x3 + 0.17x1 - 0.117x4$
7	13	0.1012	$y = 0.658x3 + 0.14x1 - x2x3x4$
8	15	0.0999	$y = x3 + 0.168x2 - x3x4 - x2x32$
9	17	0.0989	$y = x3 + 0.111x2 - x2x32 - 0.628x2x4$
10	19	0.0907	$y = x3 + 1.85x32 + 1.61x22 - 4.89x2x3$
11	21	0.0891	$y = x3 + x32 + 1.2x22 - 0.136x4 - 3.25x2x3$
12	23	0.0868	$y = x3 + 1.36x32 + 1.36x22 - 0.115x4 - 3.9x2x3$
13	25	0.0853	$y = x3 + 1.4x32 + 1.33x22 - 0.348x3x4 - 3.87x2x3$
14	35	0.0809	$y = 3.41x1x3 + 2.43x32 + 2.97x1x22 - 1.78x33 - 11.5x1x2x3$
15	37	0.0770	$y = 1.81x1x3 + 1.7x32 + 5.31x1x23 - x33 - 14.5x1x3x22$
16	39	0.0721	$y = 1.63x1x3 + 1.63x32 + 5.4x1x23 - x34 - 14.5x1x3x22$
17	41	0.0710	$y = 0.0201 + 1.57x1x3 + 1.6x32 + 5.09x1x23 - x34 - 14x1x3x22$
18	43	0.0698	$y = 0.0152 + 1.58x1x3 + 1.58x32 + 5.39x1x23 - x35 - 14.6x1x3x22$

C, complexity; MAE, mean absolute error; x1, marine environment governance; x2, marine technological progress; x3, human capital; x4, marine resource endowment.

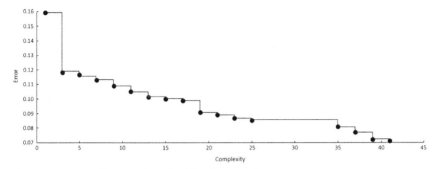

Figure 3.7 Pareto frontier described by the models.

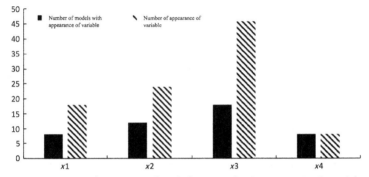

Figure 3.8 Appearance frequency of each factor in the Pareto optimal model.

As an indirect factor, marine environmental governance has relatively little impact on marine resource utilization efficiency. One reason is that China pays too much attention to economic benefits, while ignoring the importance of environmental governance in the utilization of marine resources, which makes it difficult to solve existing pollution problems and prevent future possible marine environmental problems. The other reason is that the intensity of environmental governance in China during the research period was variable, which shows that such work is not continuous, and the implementation of various environmental governance projects is often done on a "better late than never" approach.

Marine resource endowment also affects marine resource utilization efficiency, but the strength of the influence is weaker. With a change in the marine resource endowment in a region, the marine resource utilization efficiency will inevitably change, that is, the region will enjoy the "blessing of resources" due to its abundant natural assets. However, according to the "resource curse," the richer the natural resources, the lower the utilization capacity of these resources and labor in the natural resource sector. This will lead to a lack of technological progress and a reduction in resource utilization. In addition, the economic rents that each sector receives for accessing natural resources also contributes to the "voracity effect," which partially offsets the positive effects of energy reserves. For the above reasons, the impact of marine resource endowment on marine resource utilization efficiency is relatively small.

3.7 Conclusion and suggestions

3.7.1 Conclusion

By combining the superefficiency SBM-undesirable output model with the metafrontier function, the marine resource utilization efficiencies of 11 coastal areas in China are calculated, and the dynamic evolution and influencing factors of marine resource utilization efficiency in these areas are analyzed using the distribution dynamic method and symbolic regression method. The research results show that:

1. In general, regional differences in marine resource utilization efficiency under both metafrontier and group frontier are both relatively significant, and all show a decreasing pattern from east to north to south. The efficiency in the southern coastal region is far lower than the efficiencies in the northern and eastern regions, and the efficiency in the eastern region is higher than that in the northern region, no matter whether under the metafrontier or group frontier approaches. At present, full use of resources and environmental protection are more highly attained in China's eastern coastal region, where the marine economic development mode is relatively sustainable; however, the marine economy modes in the northern and southern coastal regions are relatively underdeveloped, which means these two regions should strive for comprehensively coordinated and balanced development that takes both economic and environmental benefits into consideration.

2. Regionally speaking, the regional gap in marine resource utilization efficiency presents an increasing trend that mainly comes from interregional gaps. In any case, the interregional gap can hardly surpass the intraregional gap in the short term to become the deciding force in the regional gaps in marine resource utilization efficiency. This

indicates that narrowing the gaps among areas in the eastern, southern, and northern regions should be the key to improving marine resource utilization efficiency in China. From the perspective of time series changes, the eastern region shows a slight downward trend, the northern region shows an upward trend, and the southern region shows a horizontal development trend. Under the group frontier technological level, variation trends of the three groups are close to those under the metafrontier. Generally, the marine resource utilization efficiency was in a declining trend between 2006 and 2009, on the rise between 2009 and 2011, and witnessed slight increases over the whole research period.

3. Marine resource utilization efficiency in China shows a trend of growth, but the peak height is waning, suggesting that the concentration ratio of marine resource utilization efficiency in each area is decreasing, while the dispersion is increasing. There is a twin-peak distribution that shows a trend of transforming to a single-peak structure, indicating that a polarization of marine resource utilization efficiencies exists throughout the research period, but weakening as time goes by. Meanwhile, the change spans of marine resource utilization efficiency during the whole research period and the three subperiods are all small and with poor fluidity, showing a relatively significant clustering phenomenon.

4. The order of influences of each factor on the marine resource utilization efficiency from high to low is successively: sea-related employees' scientific attainment, marine technological progress, marine environmental governance, and the marine resource endowment. Therefore on the premise of considering the real environment of marine resource utilization in each area, we should endeavor to improve the scientific attainments of sea-related employees to maintain high-level scientific activities in each area and improve the marine resource utilization efficiency. Moreover, the improvement of marine resource utilization efficiency is also inseparable from scientific and technological progress, which reflects the scientific research level of an area and its ability to transform and apply scientific research results.

3.7.2 Suggestions

1. There is still a great deal of room for improving China's marine economy, especially for the northern and southern coastal regions, where

the space for resource conservation and efficiency improvement is still significant. Therefore coastal provinces and cities should strengthen cooperation and exchanges with each other; break protectionist restrictions and administrative barriers; enhance communications on marine technology factors; attach importance to the introduction of new technology; improve the system of marine resource utilization and management; and balance the development of marine economy, ecological environment, and social development. Furthermore, the level of regional science and technological activities should be improved to exert a promoting effect on marine resource utilization efficiency.

2. As can be seen from the empirical test of marine resource utilization efficiency and its influencing factors, the scientific attainments of sea-related employees (the proportion of marine professionals in the number of sea-related employees) has a great promoting effect on the improvement of marine resource utilization efficiency of marine resources. Hence, improving the proportion of marine professionals is conducive to the full utilization of marine resources. The cultivation of marine professionals cannot be achieved without the creation and development of the relevant educational institutions. Therefore the construction and development of sea-related major courses in universities and colleges should be accelerated, the recruitment of students to sea-related majors be expanded and the continuing education of existing sea-related employees should be carried out to increase the proportion of marine professionals.

References

Aigner, D. J., & Chu, S. F. (1968). On estimating the industry production function. *The American Economic Review, 58*(4), 826−839.

Akpoko, J. G. (2007). Analysis of factors influencing adoption of intermediate farm tools and equipment among farmers in the semi-arid zone of Nigeria. *Journal of Applied Sciences, 7*(6), 796−802.

Banker, R. D., Charnes, A., & Cooper, W. W. (1984). Some models for estimating technical and scale inefficiencies in data envelopment analysis. *Management Science, 30*(9), 078−1092.

Battese, G. E., Rao, D. P., & O'Donnell, C. J. (2004). A metafrontier production function for estimation of technical efficiencies and technology gaps for firms operating under different technologies. *Journal of Productivity Analysis, 21*(1), 91−103.

Cai, W., Pacheco-Vega, A., Sen, M., & Yang, K. T. (2006). Heat transfer correlations by symbolic regression. *International Journal of Heat and Mass Transfer, 49*(23−24), 4352−4359.

Charnes, A., Cooper, W. W., & Rhodes, E. (1978). Measuring the efficiency of decision making units. *European Journal of Operational Research, 2*(6), 429−444.

Chiu, C. R., Liou, J. L., Wu, P. I., & Fang, C. L. (2012). Decomposition of the environmental inefficiency of the meta-frontier with undesirable output. *Energy Economics, 34* (5), 1392−1399.

Cui, Q., & Li, Y. (2015). An empirical study on the influencing factors of transportation carbon efficiency: Evidences from fifteen countries. *Applied Energy, 141*, 209−217.

Ding, L. L., Zhu, L., & He, G. S. (2015). Measurement and influencing factors of green total factor productivity of marine economy in China. *In Forum on Science and Technology in China, 2*(2), 72−78.

Färe, R., Grosskopf, S., & Pasurka, C. A., Jr (2007). Environmental production functions and environmental directional distance functions. *Energy, 32*(7), 1055−1066.

Fukuyama, H., & Weber, W. L. (2009). A directional slacks-based measure of technical inefficiency. *Socio-Economic Planning Sciences, 43*(4), 274−287.

Hailu, A. (2003). Nonparametric productivity analysis with undesirable outputs: Reply. *American Journal of Agricultural Economics, 85*(4), 1075−1077.

Jamnia, A. R., Mazloumzadeh, S. M., & Keikha, A. A. (2015). Estimate the technical efficiency of fishing vessels operating in Chabahar region, Southern Iran. *Journal of the Saudi Society of Agricultural Sciences, 14*(1), 26−32.

Koza, J. R. (1992). *Genetic programming: On the programming of computers by means of natural selection*. MIT Press.

Li, M. C. (2014). Predictive environment carrying capacity assessment of marine reclamation by prediction of multiple numerical model. *IERI Procedia, 8*, 101−106.

Li, J. L., Cao, K., Ding, F., Yang, W. B., Shen, G. M., & Li, Y. R. (2017). Changes in trophic-level structure of the main fish species caught by China and their relationship with fishing method. *Journal of Fishery Sciences of China, 24*(1), 109−119.

Maravelias, C. D., & Tsitsika, E. V. (2008). Economic efficiency analysis and fleet capacity assessment in Mediterranean fisheries. *Fisheries Research, 93*(1−2), 85−91.

Meeusen, W., & van Den Broeck, J. (1977). Efficiency estimation from Cobb-Douglas production functions with composed error. *International Economic Review, 18*(2), 435−444.

Produit, T., Lachance, B. N., Strano, E., & Joost, S. (2010). A network based kernel density estimator applied to Barcelona economic activities. *Lecture Notes in Computer Science, 6016*(6760), 32−45.

Quah, D. (1993). Empirical cross-section dynamics in economic growth. *European Economic Review, 37*, 426−434.

Quah, D. (1997). Empirics for growth and distribution: Stratification, polarization, and convergence clubs. *Journal of Economic Growth, 2*(1), 27−59.

Seiford, L. M., & Zhu, J. (2005). A response to comments on modeling undesirable factors in efficiency evaluation. *European Journal of Operational Research, 161*(2), 579−581.

Shi, Z. L., & Huang, Z. H. (2009). A dynamic analysis of the economic growth on the provincial level in China based on kernel density estimation method. *Economic Survey, 8*(4), 229−235.

Shorrocks, A. F. (1980). The class of additively decomposable inequality measures. *Econometrica: Journal of the Econometric Society, 48*(30), 613−625.

Shwartz, M., Burgess, J. F., Jr, & Zhu, J. (2016). A DEA based composite measure of quality and its associated data uncertainty interval for health care provider profiling and pay-for-performance. *European Journal of Operational Research, 253*(2), 489−502.

Song, M., Song, Y., An, Q., & Yu, H. (2013). Review of environmental efficiency and its influencing factors in China: 1998−2009. *Renewable and Sustainable Energy Reviews, 20*(4), 8−14.

Tavassoli, M., Faramarzi, G. R., & Saen, R. F. (2014). Efficiency and effectiveness in airline performance using a SBM-NDEA model in the presence of shared input. *Journal of Air Transport Management, 34*(1), 146−153.

Tingley, D., Pascoe, S., & Coglan, L. (2005). Factors affecting technical efficiency in fisheries: Stochastic production frontier versus data envelopment analysis approaches. *Fisheries Research, 73*(3), 363–376.

Tone, K. (2001). A slacks-based measure of efficiency in data envelopment analysis. *European Journal of Operational Research, 130*(3), 498–509.

Tone, K. (2003). Dealing with undesirable outputs in DEA: A slacks-based measure (SBM) approach. *GRIPS Research Report Series*, 2003.

Tongzon, J. (2001). Efficiency measurement of selected Australian and other international ports using data envelopment analysis. *Transportation Research Part A: Policy and Practice, 35*(2), 107–122.

Wang, L., Li, Y., & Wang, M. C. (2012). The empirical analysis of SBM efficiency in China's banking industry based on modified three-stage DEA. *Shanghai Journal of Economics, 24*(06), 3-14, 22.

Wang, L.L., 2014. *Study on total factor productivity of marine economics under environmental restraints*. Qingdao: Ocean University of China.

Wang, N., & Gai, M. (2018). Study on efficiency and effect of ecological undesirable outputs SBM model of Bohai Bay area based on factors. *Resource Development and Market, 4*(06), 741–746.

Wanke, P., Barros, C. P., & Nwaogbe, O. R. (2016). Assessing productive efficiency in Nigerian airports using fuzzy-DEA. *Transport Policy, 49*, 9–19.

Yang, G. F., Li, W. L., Wang, J. L., & Zhang, D. (2016). A comparative study on the influential factors of China's provincial energy intensity. *Energy Policy, 88*, 74–85.

Yang, L., Han, K. J., & Chen, Z. Y. (2015). Dynamic relationship between economic growth in coastal and marine disaster losses: 1989–2011. *Scientia Geographica Sinica, 35* (8), 969–975.

Yu, M. M. (2010). Assessment of airport performance using the SBM-NDEA model. *Omega, 38*(6), 440–452.

Yu, Y., & Choi, Y. (2015). Measuring environmental performance under regional heterogeneity in China: A metafrontier efficiency analysis. *Computational Economics, 46*(3), 375–388.

Yuan, F., Chen, J., & Shao, X. (2015). The evaluation on insurance e-commerce website's efficiency based on SBM-DEA. *Insurance Studies* (3), 36–45.

Zhang, B., Sun, S., & Tang, Q. S. (2013). Carbon sink by marine fishing industry. *Progress in Fishery Sciences, 34*(1), 70–74.

Zhong, Z. C. (2011). Research on international comparison of innovation SBM efficiency: Empirical analysis based on OECD members and China. *Journal of Finance & Economics, 37*(9), 80–90.

Zhou, P., Ang, B. W., & Poh, K. L. (2006). Slacks-based efficiency measures for modeling environmental performance. *Ecological Economics, 60*(1), 111–118.

CHAPTER FOUR

Analysis of the marine carbon sink capacity in China

4.1 Introduction

Industrial Revolution is undoubtedly one of the most important events in the long course of human history. While promoting the development of industrial productivity, it has also changed people's lives dramatically. However, environmental damage has increased significantly in the meantime. The burning of large amounts of coal, oil, and natural gas, and changing land-use modes have resulted in large emissions of greenhouse gases such as CO_2 and CH_4. American scholars Keeling et al. (1989) and Watson, Rodhe, and Oesehager (1990) studied changes in CO_2 concentration in the global atmosphere and found that before the industrial revolution, the CO_2 content in the air was about 280 ppm, while after the Industrial Revolution, this increased by 25%, reaching 350 ppm. Since the 20th century, the increase in global temperature has been driven by large amounts of CO_2 produced and emitted by the human population. Over the past 60 years, greenhouse gases such as CO_2 may have increased the global temperature by 0.5°C−1.3°C (Vitousek, Mooney, Lubchenco, & Melillo, 1997). If no policy measures are undertaken to reduce greenhouse gas emissions, the global average temperature may rise by 1.4°C−5.5°C by 2100. Considering the situation before the Industrial Revolution as a benchmark, every increase of 1°C will raise the sea level by 0.2−0.6 m (IPCC, 2007). Simulation using Atmosphere-Ocean General Circulation Models predicted that global sea levels will rise by an average of 0.09−0.88 m by 2100 (IPCC, 2007). The greenhouse effect destroys the natural environment and threatens the living environment of humans and the development of the social economy. Therefore interventions to reduce greenhouse gas emissions are necessary. In June 1992 at the United Nations Conference on Environment and Development, the international community reached a consensus on the current serious climate change issues. The participating countries jointly

Sustainable Marine Resource Utilization in China.
DOI: https://doi.org/10.1016/B978-0-12-819911-4.00004-7

signed the United Nations Framework Convention on Climate Change, which was the first international convention in the world to mitigate climate warming by controlling greenhouse gas emissions in a comprehensive; this has provided the basic framework for every country in the world to limit global climate change. To help countries reduce greenhouse gas emissions and reach targets quickly, participating countries signed a supplementary provision, namely the Kyoto Protocol in 1997. The purpose of this was to "ensure that greenhouse gas concentrations are maintained at an appropriate level to ensure that the ecosystem can adapt to climate change, guarantee global food security, and sustainable development of human society and economy."

Global changes have attracted the attention of the international community and have become a research hotspot in the scientific community. In many studies, researchers have focused on the global biochemical cycle of carbon, which has always been the core issue of global change and sustainable development. The total ocean area on the Earth is about 70%, which is twice as large as the total land area. Additionally, oceans are the largest reservoir of CO_2. The total carbon storage of oceans on Earth is about 3.81×10^4 Gt, which is 20 times that of the land and 50 times that of the atmosphere. Every year, about 90 Pg CO_2 circulates through oceans (González et al., 2008), and about one-third of the CO_2 released through human activity is absorbed by oceans (Khatiwala, Primeau, & Hall, 2009). The ocean is an important part of the ecosystem, and its role in the absorption and fixation of CO_2 in ecosystem services has been highlighted in the current situation of deteriorating atmospheric environments. Looking ahead, it is predicted that the absorption of CO_2 by the terrestrial biosphere may be greatly reduced due to limited nitrogen and phosphorus; however, in contrast, the absorption of CO_2 by oceans will continue to increase until 2100. Therefore the oceans will play an increasingly important role in future global CO_2 recycling system, and associated research on marine carbon sinks will be of great significance.

4.2 Literature review

To date, the study of terrestrial carbon sinks has generated a set of systematic methods, but due to the extensity and infinity of oceans, study of the marine carbon sink remains in its infancy. According to the available literature, research on marine carbon sinks has mainly focused on the primary productivity of phytoplankton and the fishery carbon sink.

4.2.1 Research on the primary productivity of marine phytoplankton

Marine phytoplankton can fix more than 36.5 Pg carbon (González et al., 2008) per year through photosynthesis. Currently, several methods are used to measure the primary productivity of phytoplankton: oxygen measurement (black and white bottle method), carbon measurement (^{14}C method), the chlorophyll method, and the remote-sensing inversion method.

Oxygen measurement (black and white bottle method) is the classic method and is based on the principle that phytoplankton releases oxygen via photosynthesis under daylight and consumes oxygen under dark conditions via respiration. The experiment is performed in two groups: a black bottle and a white bottle. The difference in oxygen released through photosynthesis in the white bottle and oxygen consumed by respiration in the black bottle represents the net oxygen production of phytoplankton. Then carbon sequestration by phytoplankton can be calculated based on each 1 mL oxygen consuming 0.536 mg carbon. In theory, this method is simple in theory; however, it has many limitations. First, zooplankton and bacteria in water can respire and release oxygen, which will increase the rate of photosynthesis. Second, the size of the operation is considerable and the sensitivity is low. It can only be applied to sea areas where the levels of photosynthesis are very strong, because the long-term exposure of black and white bottles will lead to the mass propagation of bacteria and introduce further errors into the results.

Carbon measurement (^{14}C method) is the most commonly used method. Nielsen (1952) first used this method to measure photosynthesis in phytoplankton. The method uses radioisotope ^{14}C as a quantitative tracer with high accuracy and sensitivity. The basic premise is to add a certain amount of bicarbonate or carbonate (marked by ^{14}C) to the seawater with known total CO_2 content, cultivate for a given period of time, determine the content of ^{14}C in phytoplankton, and then calculate the rate of photosynthesis in phytoplankton. The formula is as follows:

$$P = \frac{(N_S - N_b) \times C}{t \times A} \tag{4.1}$$

where P is the primary productivity of phytoplankton, N_S is the content of organic ^{14}C content in samples, N_b is the content of organic ^{14}C in blank samples, C is the total CO_2 content in seawater, t is the incubation time, and A is the quantity of ^{14}C content in samples.

Application of the ^{14}C method should satisfy the following assumptions: (1) phytoplankton absorbs and assimilates ^{12}C at the same rate as ^{14}C; (2) ^{14}C enters phytoplankton cells only through photosynthesis; (3) ^{14}C, which has been assimilated, is not excreted outside the cells; (4) the assimilated ^{14}C is not lost during the respiration of phytoplankton. However, these four assumptions may not reflect the situation in reality; thus a series of amendments must be applied to the results. In addition, as the amount of standard radioactive solution, the purity of the tracer, and the concentration of CO_2 in water samples will impact the results, the selection and accurate determination of these variables must be considered for application to the ^{14}C method.

Because photosynthesis is dependent on chlorophyll (mainly chlorophyll a), the chlorophyll method is one of the main methods used to estimate primary productivity in sea areas. Application of the chlorophyll method can be divided into three categories. The first is the Rythert−Yentsch empirical mode, as shown by formula (4.2). The assimilation coefficient Q differs under different conditions; thus it is necessary to accurately determine the assimilation coefficients reasonably and accurately under different conditions in different sea areas. The content of chlorophyll a C can be measured by spectrophotometry and fluorescence, with spectrophotometry being the most reliable.

$$P = \frac{R}{K} CQ \qquad (4.2)$$

where P is the daily photosynthetic production of phytoplankton in the unit area and its unit is $mgC/(m^2\ day)$; R is the relative photosynthetic rate under the corresponding sea-surface radiation value; K is the diffuse attenuation coefficient of seawater, of which the unit is m^{-1}; Q is the assimilation coefficient, namely the photosynthetic capacity generated by chlorophyll a in unit time under conditions of light saturation and its unit is $mgC/(mg\ chlorophyll\ a\ h)$; C is the chlorophyll a content in water and the unit is mg/m^3.

The second is the ecology mathematical statistics mode. In this mode, light transmission in seawater and the photosynthetic response of phytoplankton are calculated, and then the relationship between primary productivity and its influencing factors is expressed through mathematical relationships. Finally, productivity is calculated based on known data of the influencing factors.

The third is the remote-sensing mode. Clarke, Ewing, and Lorenzen (1970) were the first to use remote-sensing technology to measure the chlorophyll concentration in marine phytoplankton. Through satellite observations and inversion, this mode can estimate marine primary productivity on a large scale, which is an advantage compared with other methods, and will become a focus of future bio-oceanographic research. Inversion of phytoplankton primary productivity based on the remote-sensing mode can be divided into two types: the inversion of chlorophyll and primary productivity through the relationship between chlorophyll concentration and primary productivity, and the direct inversion of primary productivity.

To date, scholars have studied the primary productivity of marine phytoplankton. Marine phytoplankton captures more than 36.5 Pg C/year through photosynthesis (Yang et al., 2005). Teira et al. (2005) investigated primary productivity in the eastern part of the North Atlantic subtropical cyclone zone from 1992 to 2001. They found that microphytoplankton ($<2\,\mu m$) accounted for the majority of phytoplankton (71%) and its primary productivity accounted for 54% of the total primary productivity. Seeyave, Lucas, Moore, and Poulton (2007) studied phytoplankton productivity in the Southern Ocean from 2004 to 2005; primary productivity estimated from remote-sensing chlorophyll data revealed a distinct feature of the north−south gradient. Zhang, Ding, Li, Xue, and Guo (2016) measured the primary productivity of the East China Sea during four seasons from 2008 to 2009 by sea and found that the average primary productivity of the four voyages was 375.03, 414.37, 245.45, and 102.60 mg/(m^3 h), respectively, summer > autumn > spring > winter. Lv, Xia, Li, and Fei (1999) studied changes in chlorophyll a and primary productivity in Bohai Sea during 1982−83 and 1992−93 using the ^{14}C method and showed that chlorophyll a decreased from 1.05 to 0.61 mg/m^3, and primary productivity decreased from 312 to 216 mg/m^2 day; the spatial distributions of chlorophyll a and primary productivity were consistent with features of seasonal variations. Tin, Lomas, and Ishizaka (2015) generated chlorophyll a concentration and photosynthetic effective irradiance curves by remote-sensing inversion and estimated the primary productivity of phytoplankton in the Atlantic Sargasso Sea from 2004 to 2009. Tilstone et al. (2015) compared the ocean color remote-sensing model with the physical−biological coupling model and estimated the primary productivity of the Arctic Ocean by combining the two models. Hyde, O'Reilly, and Oviatt (2008) performed remote-sensing inversion of the primary productivity of Massachusetts Bay (United States) using the VGPM model and found that changes in primary productivity over time were well reflected by this model, and the spatial and instantaneous resolutions of the estimated results were significantly improved.

4.2.2 Research status of carbon sink fisheries

The concept of a carbon sink fishery was first proposed by Qisheng Tang in 2010. As fishery production activities can cause aquatic organisms to absorb CO_2 from water, and carbon can be removed from water by harvesting aquatic products, improving the ability of aquatic ecosystems to absorb CO_2 in air, a "carbon sink fishery" is defined as a fishery activity that leads to the function of a carbon sink and directly or indirectly reduces the effect of CO_2. Carbon sink fishery is a fishery production activity that does not involve the use of bait, which can be divided into two main categories: one is the cultivation of shellfish and algae. Filter-feeding bivalves mainly feed on phytoplankton and have strong water-filtering ability (Bacher, Bioteau, & Chapelie, 1995; Riisgard, 1991) they produce a strong "top-down control effect" that influences the number of phytoplankton (Wong et al., 2003), reduces the concentration of suspended particulate matter in water (Cerrato, Caron, Lonsdale, Rose, & Schaffner, 2004), and increases the transparency and light depth of water, which is beneficial for the growth of benthic plants (Peterson & Heck, 1999). Therefore the capacity to absorb and fix carbon is strengthened (Lee & Kenneth, 1997) and the carbon sink potential of water is improved. In addition, the formation of shells (mainly $CaCO_3$) is a process of carbon immobilization in water. Like phytoplankton, macroalgae as primary producers can absorb CO_2 from water through photosynthesis and store it in organisms through assimilation. By harvesting, cultured shellfish and algae can transfer biological carbon sinks from water. The other is the culture of predatory fish that feed on plankton, shellfish, and algae. Through carbon transfer of the food chain, part of the fixed carbon of plankton and shellfish is transferred to the body of predatory fish to maintain growth and activity, and movable carbon sinks can be formed through capture. Furthermore, carbon sink fishery also includes proliferation and release, artificial reefs, and fishing fisheries. To date, most research on fishery carbon sinks has been performed to calculate the carbon sinks of cultured shellfish and algae. Academics have reached a consensus on the method of calculation, to estimate the carbon sinks of cultured shellfish and algae using data on shellfish and algae production and carbon content based on the conservation of carbon elements in the process of material transfer.

Scholars have performed much research on fishery carbon sink. Liu, Ren, and Zhang (2011) estimated that shellfish and algae cultivation in the Bohai Rim in 2009 utilized more than 1.8 million tons of carbon from marine ecosystems and removed 574.2 thousand tons of carbon from seawater. Therefore carbon sink fishery based on shellfish and algae culture has significant economic, ecological, and social benefits. Zhang and Choi (2013) used the ecological conversion efficiency of each trophic level and determined the total carbon sequestration of the fishing industry in the Yellow Sea and Bohai Sea from the catches. The annual carbon sequestration of the fishing industry in the Bohai Sea and Yellow Sea from 1980 to 2000 was 2.83−10.08 and 3.61−26.13 million tC, respectively. Alpert, Spencer, and Hidy (1992) predicted the carbon sequestration potential of macroalgae in the continental shelf area and reported that this was about 0.7 GtC/a globally, which accounted for about 35% of the global annual average net carbon sequestration (2.0 ± 0.8 GtC/a), under the cultivation cost of 300 \$/tC a. Whittaker (1975) estimated the net primary productivity of the oceanic zones, continental shelf areas, and estuaries and algae areas and calculated that the average net primary productivity of the oceanic zones was about 125 gC/m^2 a, while that of the continental shelf areas was about 360 gC/m^2 a, and that of estuaries and algae areas were as high as 2500 gC/m^2 a. In addition, Chapman (1974), Macroy and McMillan (1977), and Mann and Chapman (1975) calculated the carbon sequestration of different algae based on the data exchange algorithm of seaweed.

In summary, existing studies start from the sea to research the conditions of the carbon sink, and no studies have investigated the marine carbon sink in coastal provinces and cities. This chapter will first consider how to estimate the carbon sink capacity of the ocean for carbon sequestration. Then considering the overall situation of the marine carbon sink, carbon sequestration via a "biological pump" and a "carbonate pump" are summed to provide a method of measuring the marine carbon sink. Carbon sequestration by "biological pump" can be divided into phytoplankton carbon sequestration and macroalgae carbon sequestration. Carbon sequestration by "carbonate pump" is mainly via shellfish carbon sequestration. Then empirical analysis of the marine carbon sink conditions of 11 coastal provinces and cities in China from 2008 to 2015 is performed to investigate their temporal and spatial evolution, and the Theil index method is used to analyze the regional differences in the marine carbon sink.

4.3 Concept of marine carbon sink and formation

4.3.1 Concept of marine carbon sink

4.3.1.1 Net marine carbon sink

In the United Nations Framework Convention on Climate Change, the process of releasing greenhouse gases into the atmosphere is defined as the "source" of greenhouse gases, while the process of removing greenhouse gases from the atmosphere is defined as the "sink" of greenhouse gases. For the whole marine ecosystem, the ability to absorb and release CO_2 is compared. The sea area that can absorb CO_2 from the atmosphere is termed the "net carbon sink," and the sea area that can release CO_2 into the atmosphere is termed the "net carbon source" (Fig. 4.1).

Two concepts, namely the "Apparent source and sink" and the "Realistic source and sink," should be distinguished here. The "Apparent source and sink" is derived from the positive and negative CO_2 fluxes at the air−sea interface and is determined by measuring the partial pressure of CO_2 in the surface seawater and atmosphere. The formula is as follows:

$$F = K\left(p_{CO_2,air} - p_{CO_2,sea}\right) \tag{4.3}$$

where F is the carbon flux at the air−sea interface (g/m^2 a); K is the gas exchange coefficient; $p_{CO_2,sea}$ is the partial pressure of CO_2 in surface seawater (Pa), and $p_{CO_2,air}$ is the partial pressure of CO_2 in the atmosphere (Pa). The "Realistic source and sink" refers to the source and sink obtained by comparing total CO_2 in current seawater with that in seawater before the Industrial Revolution. The "Apparent source and sink" is usually suitable for small-scale waters with spatial and temporal heterogeneity. The "Realistic source and sink" is generally applicable to the acquisition of marine carbon sources and sinks on an Ocean scale, which can reflect the carbon storage capacity of the middle and deep seas. As the middle and deep seas are relatively stable, the long-term and large-scale relationship between ocean and atmospheric CO_2 can be reflected, and trends in the intensity and variation of marine carbon sources and sinks can be forecast through model inversion.

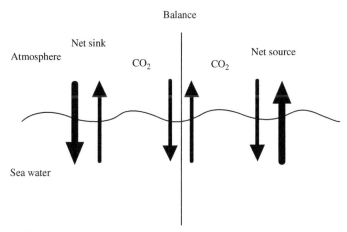

Figure 4.1 Net marine carbon source and sink.

4.3.1.2 Marine carbon sink

The net carbon source and sink of the ocean can be divided into two processes, that is, the source process of releasing CO_2 into the atmosphere and the sink process of absorbing CO_2 from the atmosphere. Carbon sources and sinks are found in any sea area, and the algebraic sum of carbon sequestration and carbon emission is the net carbon source and sink of the ocean. As a carbon source, oceans release carbon into the atmosphere from land-derived inputs, atmospheric deposition, and the respiration of marine organisms, while carbon in the marine sink is mainly that sequestrated by marine organisms originating from the dissolution of CO_2 at the air—sea interface. Phytoplankton fixes CO_2 through photosynthesis, which is then transmitted to various marine organisms through the food chain and food web to achieve CO_2 fixation.

4.3.2 Carbon cycle in the ocean

Carbon is the key component of organic life on Earth and plays an important role in marine and terrestrial biogeochemical cycles. Carbon in the oceans can be classified into three categories according to its sources: carbon generated by marine ecosystems, by terrestrial ecosystems, and by human activities. There are three main forms of carbon in the ocean: dissolved inorganic carbon (DIC), dissolved organic carbon (DOC), and particulate organic carbon (POC) (Committee on Global Change, 1988; Eatherall, Naden, & Cooper, 1998). Fig. 4.2 illustrates the marine carbon cycle.

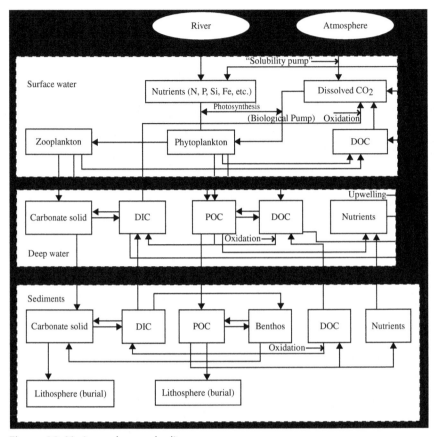

Figure 4.2 Marine carbon cycle diagram.
Note: *DIC*, Dissolved inorganic carbon; *DOC*, dissolved organic carbon; *POC*, particulate organic carbon (Yang, Li, & Pan, 2004).

The equilibrium relationships among each form of inorganic carbon in the seawater CO_2 system are as follows:

$$CO_2\ (g) \rightleftharpoons CO_2\ (aq) \tag{4.i}$$

$$CO_2\ (aq) + H_2O \rightleftharpoons H_2CO_3 \tag{4.ii}$$

$$H_2CO_3 \rightleftharpoons H^+ + HCO_3^- \tag{4.iii}$$

$$HCO_3^- \rightleftharpoons H^+ + CO_3^{2-} \tag{4.iv}$$

$$Ca^{2+} + 2HCO_3^- \rightleftharpoons CaCO_3(s) + CO_2\ (g) + H_2O \tag{4.v}$$

During its dissolution in seawater, CO_2 interacts with water molecules. Therefore four main forms of inorganic carbon exist in seawater: HCO_3^-,

CO_3^{2-}, CO_2, and H_2CO_3 (CO_2^* is commonly referred to as dissolved CO_2 and H_2CO_3 is known as free CO_2). DIC accounts for more than 98% of total seawater carbon (Zeebe & Wolf-Gladrow, 2001), in which HCO_3^- is the main component that accounts for more than 90%, followed by CO_3^{2-} accounting for about 9%. The remaining <1% is dissolved CO_2 and H_2CO_3 (Bakker, de Baar, & de Jong, 1999). These values are only estimates, and the composition of various forms of carbon varies in different sea areas and under different hydrology and pH conditions. In addition, HCO_3^- and CO_3^{2-} are important for the alkalinity of seawater (ALK), accounting for more than 95% of the alkalinity content (Gong, Zhang, & Zhang, 2006).

Organic substances in the ocean include particle organic matter (debris), dissolved organic matter, phytoplankton, zooplankton, and bacteria. The content of organic matter in seawater is usually expressed as the content of carbon. Organic carbon refers to the carbon in organic molecules linked to other atoms by chemical bonds. Dissolved organic carbon is distinguished from POC by the ability to pass through a 0.45-μM filter membrane. Dissolved organic carbon accounts for about 90% of the total organic matter in ocean water, and POC accounts for about 10%. The main internal sources of dissolved organic carbon in seawater are marine biological and chemical processes, which are fundamentally derived from photosynthesis of phytoplankton. Among the carbon fixed by phytoplankton through photosynthesis each year, the carbon oxidized by the respiration of marine organism is estimated to account for about 99.95%, and only about 0.05% avoids oxidation and is buried in sediments.

4.3.3 Mechanism of CO_2 absorption in the ocean

CO_2 is absorbed by the ocean via four mechanisms: "solubility pump," "biological pump," "microbial carbon pump," and "carbonate pump." "Solubility pump" is the chemical equilibrium and physical transfer process of CO_2 at the air—sea interface. Due to differences in seawater temperature, salinity, ocean circulation, and CO_2 concentration between air and sea, CO_2 in the atmosphere can enter the water body through dissolution and transform into DIC such as bicarbonate and carbonate. At the same time, CO_2 in the ocean can be released into the atmosphere in molecular form after saturation. Atmospheric CO_2 can be absorbed by seawater at high latitudes and then transported to the equator via the action of physical pumps. The efficiency of the solubility pump depends on the latitude and seasonal variations of the ocean's thermohaline circulation and ocean currents. Sea areas of low temperature and high salinity are more conducive to CO_2 dissolution.

The biological pump transports organic carbon and is also known as the "organic carbon pump." First, autotrophic phytoplankton in the ocean absorb dissolved CO_2 from seawater by photosynthesis; this converts DIC into POC, resulting in marine primary productivity. POC is continuously transferred and circulated via transmission of the food chain and food web. Through the processes of respiration, excretion, death, and decomposition, marine organisms at all levels discharge a large amount of biodetritus into seawater in the form of inanimate POC and produce a large amount of inert dissolved organic carbon in the process. Microorganisms can decompose most POC into inorganic carbon, making it reenter the ocean carbon cycle, while the remaining POC is deposited on the seabed, where it is temporarily separated from the carbon cycle (inert dissolved organic carbon is not directly involved in the ocean carbon cycle), remaining in the ocean for tens to thousands of years (Mikaloff-Fletcher, Gruber, Jacobson, & Doney, 2006; Quay, Sonnerup, Westby, Stutsman, & McNichol, 2003). This process of formation, consumption, and transport of organic matter, and POC penetrating deep into the ocean is known as the "biological pump," which originates from the photosynthesis of phytoplankton (Chisholm, 2000; Falkowski, Barber, & Smetacek, 1998). Most of the carbon transported by the biological pump is POC, and only a small amount is dissolved organic carbon. However, POC accounts for more than 95% of total organic carbon (Cauwet, 1977) in the ocean, most of which cannot be used by organisms; therefore this part of carbon is fixed in the ocean over the long term.

The microbial carbon pump is a form of "biological pump," and this concept was first proposed by Prof. Nianzhi Jiao in 2010. This mainly refers to the process through which marine microorganisms modify and transform dissolved organic carbon to inert dissolved organic carbon through a series of physical and chemical processes (inert dissolved organic carbon includes prions, peptidoglycans, polyester polysaccharides, and some D-amino acids). As inert dissolved organic carbon cannot be used by most organisms, it can be stored in the ocean over the long term. The total amount is about 650 Gt, which is equivalent to the atmospheric carbon pool. Therefore the microbial carbon pump plays an important role in carbon sequestration in the ocean. Particularly, the microbial carbon pump plays an essential role in estuaries and shallow sea areas where the biological pump is substantially weakened due to resuspension. In some areas of the South China Sea, the primary productivity contributed by microphytoplankton and supermicrophytoplankton is estimated at around 60% of the total.

The carbonate pump refers to the process through which shellfish, coral, and other organisms in the ocean use bicarbonate (HCO_3^-) in seawater to produce a calcium carbonate ($CaCO_3$) shell (e.g., shell, coral calcium). $CaCO_3$ formation involves the absorption of carbon elements in seawater, which belong to the "sink" of CO_2. At the same time, CO_2 is released, which is the "source" of CO_2. The rainfall ratio (rainfall = PIC/POC = $CaCO_3/C_{org}$) is usually used to indicate the relative strength of the "source" and "sink." Only when the rainfall rate exceeds 1.5 can the "carbonate pump" be regarded as a net carbon sink. The formation and dissolution of $CaCO_3$ can affect the acidity and alkalinity of the ocean and may be important for the changing global atmospheric CO_2 concentration.

4.4 Introduction and construction of marine carbon sink estimation methods

In this section, the term "carbon sink" refers specifically to the ability of the ocean to absorb and fix CO_2, regardless of the impact of the "carbon source." According to the literature, the total carbon reserve in the ocean is 3.81×10^4 GtC/a, whereby the total carbon sequestration of marine organism is 3 GtC/a and the net carbon absorption during CO_2 dissolution at the air—sea interface is 1.6 GtC/a (IPCC, 2007). Therefore the carbon sequestration capacity of the ocean can be characterized by the estimated carbon storage capacity of marine organisms. Hence, in this chapter, the carbon sequestration capacity of the ocean is estimated by summing the "biological pump" and the "carbonate pump." Although CO_2 is released during the formation of $CaCO_3$, as only 1 mol CO_2 is released for absorption of every 2 mol carbon and the other 1 mol is stored in the shells of shells, meaning 1 mol carbon in the seawater is absorbed during the formation of shellfish and the carbon content in seawater is reduced, it is reasonable to add the influence of "carbonate pump" in the calculation. Carbon sequestered by the "biological pump" in the ocean is mainly derived from the photosynthesis of phytoplankton and macroalgae as primary producers, while the "carbonate pump" is mainly derived from the shell formation of shellfish. Therefore three parts of the marine carbon sink can be measured: carbon sequestration by phytoplankton, macroalgae, and shellfish.

4.4.1 Estimation of carbon sequestration by phytoplankton

Photosynthetic organisms in the ocean can be divided into five main categories: phytoplankton, macroalgae, marine angiosperms, photosynthetic bacteria, and benthic diatoms. Phytoplankton are the microsingle-celled algae that can freely suspend in water. They are widely distributed, found in large quantities in the ocean, and are the main producer in the marine ecosystem. As the primary productivity of phytoplankton can account for 99% of the primary productivity of the whole ocean, they are of great significance for estimating the carbon sink of the whole ocean.

At present, the main methods used to measure the primary productivity of phytoplankton are oxygen measurement (black and white bottle method), carbon measurement (^{14}C method), chlorophyll method, and remote-sensing method. Because these methods require measured data, and data acquisition is difficult, this chapter will refer to the methods of Zhang and Choi (2013) for estimating phytoplankton carbon sequestration in fishing industry by catches (as shown in Fig. 4.3), and the method of Ye and Tang (1980) for estimating the herring population in the Yellow Sea using fishing mortality to measure carbon sequestration of marine phytoplankton. First, data on the marine fishing volume in coastal areas are obtained from the *China Fishery Statistical Yearbook* (as algae are also primary producers and their carbon sequestration is involved in Section 4.4.2, to avoid repeated calculation, algae catch should be deducted here). The ratio of fishing volume to fishing mortality is the stock of resources, and the transfer mechanism of carbon elements in the food chain is used to deduce the carbon content fixed by phytoplankton as a primary producer. The specific calculations are as follows:

1. Estimation of aquatic resources stock

First, the resource stock can be estimated by the ratio of catches to fishing mortality, that is

$$N_0 = \frac{Y}{\overline{F}} \tag{4.4}$$

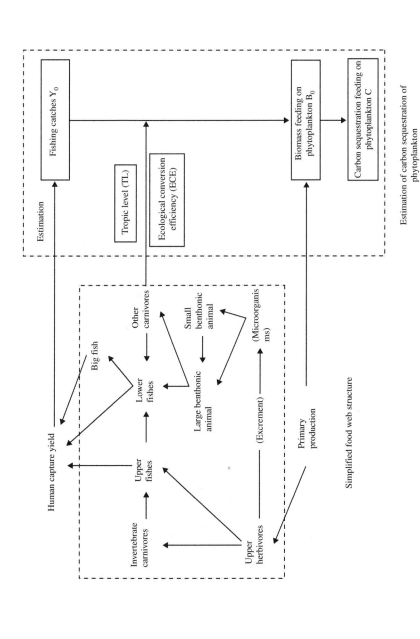

Figure 4.3 Simplified food web structure and estimation of phytoplankton carbon sequestration (Zhang & Choi, 2013).

where N_0 is amount of resources, Y is the difference in total marine catches subtracting the catches of algae, and $\overline{F}$ is the mortality rate of fishing.

The mortality rate of fishing cannot be measured separately. Generally, the total mortality rate is first measured and then separated into natural mortality rate and fishing mortality rate. The population decline can be expressed by the following two formulas:

$$\frac{dN}{dt} = -(F + M)N \tag{4.5}$$

$$N_t = N_0 e^{-(F+M)t} \tag{4.6}$$

where F and M are the coefficients of fishing mortality and natural mortality, respectively.

Sampling time is set as the unit time, that is, $t = 1$. There is

$$F + M = \ln \frac{N_0}{N_t} \tag{4.7}$$

Thus the logarithm of the ratio of resource stocks in two periods is equal to the total death coefficient in value. The ratio of resource stocks can be estimated by replacing the catch per unit fishing force in two periods. The catch per unit fishing force is defined as the average yield per 100 nets of the locomotive-boat trawl, and the total death coefficient Z can be obtained.

The total death coefficient Z is equal to the sum of the fishing mortality coefficient F and the natural mortality coefficient M. The fishing mortality coefficient is proportional to the fishing force. Thus

$$Z = C'f + M \tag{4.8}$$

where f is the fishing force and C' is the proportional coefficient. The above formula can be regarded as a function of Z around f to obtain (f, Z) in different periods. Then a straight line is fitted and its intercept is the natural mortality coefficient M.

From the difference between the total mortality coefficient Z and the natural mortality coefficient M, the fishing mortality coefficient F can be obtained, and the fishing mortality coefficient can be converted to the fishing mortality rate using the following formula:

$$E = \frac{(1 - s)F}{Z} \tag{4.9}$$

where E is the fishing mortality rate and s is the survival rate.

2. Estimation of carbon sequestration by phytoplankton

First, the biomass of the next trophic level preyed by the catches is estimated. In China's offshore waters, the average trophic level of catches is between three and four; thus the following formula is used to estimate the biomass B_1 of preys between the average trophic levels one and two:

$$B_1 = \frac{Y_0}{\left(ECE_{(TL_0 - 1)} \times ECE_{(TL_0 - 2)}\right)} \qquad (4.10)$$

where Y_0 is the estimated resource stock of catches; TL_0 is the average trophic level of catches, and ECE is the ecological conversion efficiency between the two trophic levels $(TL_0 - 1)$ and $(TL_0 - 2)$.

Second, the biomass of phytoplankton and organic debris ingested by the catches is estimated. The proportion, P, of phytoplankton to organic debris (tropic level $= 1$) and primary consumers (trophic level $= 2$) are estimated according to the trophic level formula:

$$1 \times P + 2 \times (100\% - P) = TL_0 - 2 \qquad (4.11)$$

Namely,

$$P = 4 - TL_0 \qquad (4.12)$$

Then the biomass, B_0, of phytoplankton and organic debris ingested is further estimated.

$$B_0 = (B_1 \times P) + B_1 \times \frac{(100\% - P)}{ECE_{TL=1}} \qquad (4.13)$$

where $ECE_{TL=1}$ is the ecological conversion efficiency of phytoplankton and organic debris preyed on by primary consumers.

Finally, the primary productivity of phytoplankton is estimated based on the biomass of phytoplankton and organic debris preyed. The organic debris in the ocean is formed by phytoplankton deposition, marine organisms (e.g., zooplankton and fish) fecal ball packing deposition, dead organism deposition, and the formation of sea snow by the transparent exopolymer particle condensation network. The initial source of organic debris is POC generated by phytoplankton through photosynthesis. Therefore carbon sequestration by organic debris and phytoplankton is the same. The average carbon content of 10 phytoplankton species, 4.49%, was used to estimate the existing carbon content of phytoplankton, C_1, and then the carbon sequestration of preyed phytoplankton was estimated, with the carbon sequestration by phytoplankton being 45 times that of carbon content of phytoplankton.

$$C_1 = 4.49\% \times B_1 \qquad (4.14)$$

$$C = 45 \times C_1 \qquad (4.15)$$

4.4.2 Estimation of carbon sequestration by algae

In addition to phytoplankton, algae are important producers in the marine ecosystem. With the help of photosynthesis and nutrients in water, algae can possess high productivity. The carbon sequestration capacity of algae has been estimated through the production and carbon content of cultured algae; however, algae harvest cannot fully characterize the carbon sequestration capacity, as part of its fixed carbon will migrate to water or sediment through complex biochemical processes and in the form of POC and dissolved organic carbon. Most studies have shown that the release of dissolved organic carbon from macroalgae through leaves accounts for less than 5% of the total carbon sequestration. POC is mainly released by dynamic erosion and debris deposition during the harvest period. Takashi, Ichiro, and Ken (2001) studied the release of POC from Otsuchi Bay in Japan and showed that its release accounted for 19% of the total photosynthetic productivity. Therefore the following formula can be used to characterize the carbon sequestration capacity of algae photosynthesis:

$$TC_{fix} = C_{bio} + DOC_{rel} + POC_{rel} \qquad (4.16)$$

where TC_{fix} is the total carbon sequestration by photosynthesis, C_{bio} is the amount of carbon sequestration removed from seawater due to algae harvest, DOC_{rel} is the amount of soluble organic carbon released by algae, and POC_{rel} is the amount of POC and biodebris carbon produced by algae.

C_{bio} can be easily ascertained using the above formula. Data on the aquaculture and fishing yields of algae can be obtained from the *China Fishery Statistical Yearbook*, and their sum can be taken as the total algae yield (because not all algae are captured, stock data should be considered here; however, as algae carbon sequestration accounts for only a small proportion in the marine carbon sink and statistical data are insufficient, this is only roughly estimated based on yield). Then a ratio of 5:1 is used to convert wet weight to dry weight. The result is multiplied by the carbon content of algae to obtain the carbon sequestration removed by harvesting. The formula is as follows:

$$C_{bio} = P \times 20\% \times w_c \qquad (4.17)$$

where P is the algae production, that is, the sum of algae catches and cultivation, w_c is the carbon content.

It is difficult to determine the values of DOC_{rel} and POC_{rel} directly. Therefore based on previous research, their proportion in the total photosynthetic carbon sequestered and C_{bio} can be used to deduce TC_{fix}. The formula is as follows:

$$TC_{fix} = \frac{C_{bio}}{(1 - \alpha - \beta)} \qquad (4.18)$$

where α is the proportion of DOC_{rel} in the total carbon sequestered by photosynthesis and β is the proportion of POC_{rel} in the total carbon sequestered by photosynthesis. In the calculation, α is 5% and β is 19%.

4.4.3 Estimation of carbon sequestration by shellfish

The following method is used to estimate shellfish carbon sequestration. First, data on cultured and captured shellfish are obtained from the *China Fishery Statistics Yearbook*, and their sum is taken as the total output of shellfish (as shellfish carbon sequestration accounts for only a small proportion of the marine carbon sink and the total stock of marine shellfish cannot be acquired, only the output data are used here). Then the dry weight of shellfish is obtained by multiplying the dry shell weight coefficient (defined as the ratio of dry shell weight to total wet shell weight) and the carbon sequestration of shells is obtained by multiplying the dry weight of shellfish by the carbon content of shells. That is:

$$\text{Shell carbon sequestration} = \text{shellfish yield}$$
$$\times \text{dry shell weight coefficient}$$
$$\times \text{shell carbon content} \qquad (4.19)$$

4.5 Calculation of the marine carbon sink and analysis

4.5.1 Data sources and statistical description

The data used in this section were obtained from the *China Fishery Statistics Yearbook* and the China Ocean Statistics Yearbook each year from 2008 to 2015.

Taking 2008 as an example, data were processed as follows.

4.5.1.1 Estimation of primary productivity of phytoplankton

Table 4.1 shows the total amount of marine and algae catches in 11 coastal provinces and cities in 2008. To avoid repeat calculations, the ratio of total catches deducted by the algae catches to the estimated fishing mortality of 0.48 was used as the estimated value of the catch stock. Then the carbon sequestration of phytoplankton was estimated by formulas (4.10−4.15).

4.5.1.2 Estimation of carbon sequestration by algae

Table 4.2 lists the volumes of algae catches and algae cultivation in 11 coastal areas in 2008. Carbon sequestration of algae was calculated by formulas (4.16−4.18). The percentage carbon contents of cultured kelp, *Undaria pinnatifida*, porphyra, and gracilaria are shown in Table 4.7, and those of other cultured and caught algae are calculated using the mean

Table 4.1 Total marine and algae catches in 11 coastal areas in 2008.

Area	Marine catches (ton)	Algae catches (ton)
Tianjin	18,777	
Hebei	253,300	
Liaoning	1,028,217	154
Shanghai	20,055	
Jiangsu	566,308	1034
Zhejiang	2,343,219	1980
Fujian	1,833,728	2734
Shandong	2,383,213	8115
Guangdong	1,454,640	9729
Guangxi	656,024	
Hainan	938,789	12,847

Note: Missing data are not given in the fishery statistical yearbook; 0 was used in the calculation.

Table 4.2 Volume of algae catches and algae cultivation in 11 coastal areas in 2008.

Area	Algae catches (ton)	Algae culture volume (ton)					
		Total	Kelp	Undaria pinnatifida	Porphyra	Gracilaria	Others
Tianjin							
Hebei							
Liaoning	154	246,387	135,577	104,234			6576
Shanghai							
Jiangsu	1034	25,269	3263		21,927	36	43
Zhejiang	1980	32,929	5452		18,732	317	8428
Fujian	2734	509,900	419,028	2	34,475	52,992	3403
Shandong	8115	502,433	233,148	27,760	4468	3980	233,077
Guangdong	9729	50,327	1283	25	1864	45,296	1859
Guangxi							
Hainan	12,847	18,777				11,825	6952

Note: Missing data in the table are not given in the fishery statistical yearbook; and 0 was used in the calculation.
Source: Extracted from China Fishery Statistical Yearbook (2008).

carbon content percentages of the four cultured algae mentioned above. The carbon sequestration of algae is the sum of carbon sequestration of caught and cultivated algae.

4.5.1.3 Estimation of carbon sequestration by shellfish shells

Table 4.3 shows the shellfish catches and cultivation volumes by species in 11 coastal areas in 2008. Carbon sequestration by shellfish shells can be calculated using formula (4.19). The dry shell weight coefficients and carbon contents of cultivated oysters, mussels, scallops, and clams are shown in Table 4.4. The dry shell weight coefficients and carbon contents of other cultivated species and caught shellfish are calculated based on the mean values for the four kinds of shellfish. Carbon sequestration by shellfish shells is calculated as the sum of shell carbon sequestrations by caught and cultivated shellfish.

In addition, when the Theil index is used to analyze regional differences in marine carbon sinks, the gross ocean product of coastal areas, which is derived from the *China Ocean Statistical Yearbook*, is also used. Table 4.4 lists the gross ocean products of 11 coastal areas in 2008, which can be used directly in calculations.

4.5.2 Calculation

The calculations are described in 4.4. When estimating phytoplankton carbon sequestration, the catches should be calculated as the value of marine catches minus the algae catches in the Statistical Yearbook. In addition, published literature was consulted when calculating fishing mortality due to the declining population and inaccessibility of catches per unit fishing power (Leng, He, & Wei, 1984; Wu, Lou, & Zhao et al., 2011; Ye & Tang, 1980); the fishing mortality was set as 0.48.

To estimate trophic level and eco-conversion efficiency, the following method is proposed. First, according to Zhang, Tang, and Jin (2007), the average trophic level of marine catches in China has decreased in recent years. For the Yellow Sea, the average trophic level of catches decreases by 0.14 every 10 years, while for the Bohai Sea, it decreases by 0.17 every 10 years. Thus we can calculate the trophic level of catches in each year. Based on the previous studies (Li, Hao, & Chi, 2017; Liu, Mao, Ren, & Li, 2014), the trophic levels of catches in the East China Sea and South China Sea in 2015 were estimated as 3.125 and 3.1, respectively. According to Chao, Quan, Li, and Chen (2005), the average trophic level of catches in the East China Sea was 3.5 in 1965 and 2.8 in 1990, from

Table 4.3 Shellfish catches and cultivation volumes by species in 11 coastal areas in 2008.

Area	Shellfish catches (ton)	Shellfish farming volume (ton)					
		Total	Oysters	Mussel	Scallop	Clam	Others
Tianjin	4522						
Hebei	19,117	259,711	3467	1840	131,122	66,309	56,973
Liaoning	123,619	1,572,050	143,545	52,687	289,384	658,754	427,680
Shanghai							
Jiangsu	40,070	563,426	6182	20,989		349,405	186,850
Zhejiang	10,073	667,560	111,229	68,081	2203	65,847	420,200
Fujian	50,275	2,049,522	1,419,083	62,713	10,122	265,318	292,286
Shandong	254,093	2,747,401	498,419	152,655	647,161	1,176,170	272,996
Guangdong	58,790	1,606,738	825,102	112,633	56,296	274,670	338,037
Guangxi	62,858	598,577	345,957	8724	706	192,291	50,899
Hainan	20,342	15,916	1398		45	9309	5164

Note: Missing data in the table are not given in the fishery statistical yearbook; 0 is used in the calculation.

Table 4.4 Gross ocean product of 11 coastal areas in 2008.

Area	Gross ocean product/CNY100 million
Tianjin	1888.7
Hebei	1396.6
Liaoning	2074.4
Shanghai	4792.5
Jiangsu	2114.5
Zhejiang	2677.0
Fujian	2688.2
Shandong	5346.3
Guangdong	5825.5
Guangxi	398.4
Hainan	429.6

which the catches in the East China Sea and the South China Sea from 2005 to 2014 can be reverse-deduced. Lindeman efficiency (or 10% law) is not accurate for estimating the efficiency of ecological conversion, since the efficiency of energy conversion varies in the marine ecosystem due to different trophic levels and species (Pauly, Christensen, Dalsgaard, Froese, & Francisco, 1998). Therefore the formula proposed by Tang, Guo, Sun, and Zhang (2007) was used for the calculation, as follows:

$$ECE = -15.615TL + 86.235 \tag{4.20}$$

Tables 4.5 and 4.6 show the average trophic levels and ecological conversion efficiencies of catches in the four major sea areas of China in each year. For Liaoning province and Shandong province, as the waters under their jurisdiction cover both the Bohai Sea and Yellow Sea, they are calculated separately according to their corresponding sea area ratios and then summed.

Table 4.7 shows the carbon contents of four common algae. For the carbon contents of other algae, the average value for the four algae, 29.42%, is used.

Table 4.8 lists the dry shell weight coefficients and carbon contents of four common shellfish. The average values of the four shellfish were used to estimate the dry shell weight coefficients and carbon contents of other shellfish, at 59.02% and 11.76%, respectively.

4.5.3 Result analysis and discussion

4.5.3.1 Spatial and temporal evolution of marine carbon sink

The marine carbon sinks of 11 coastal areas in China from 2008 to 2015 were calculated using the described method. The results are shown in Table 4.9.

Table 4.5 Average trophic levels and ecological conversion efficiencies of catches in the Bohai and Yellow Seas.

Year	The Bohai Sea				The Yellow Sea			
	TL_0	ECE_{TL0-1}	ECE_{TL0-2}	$ECE_{TL=1}$	TL_0	ECE_{TL0-1}	ECE_{TL0-2}	$ECE_{TL=1}$
2008	3.254	0.5104	0.6665	0.7062	3.288	0.5051	0.6612	0.7062
2009	3.237	0.5130	0.6692	0.7062	3.274	0.5073	0.6634	0.7062
2010	3.220	0.5157	0.6718	0.7062	3.260	0.5095	0.6656	0.7062
2011	3.203	0.5184	0.6745	0.7062	3.246	0.5116	0.6678	0.7062
2012	3.186	0.5210	0.6772	0.7062	3.232	0.5138	0.6700	0.7062
2013	3.169	0.5237	0.6798	0.7062	3.218	0.5160	0.6722	0.7062
2014	3.152	0.5263	0.6825	0.7062	3.204	0.5182	0.6743	0.7062
2015	3.135	0.5290	0.6851	0.7062	3.190	0.5204	0.6765	0.7062

Table 4.6 Average trophic levels and ecological conversion efficiencies of catches in the East China and South China Seas.

Year	The East China Sea				The South China Sea			
	TL_0	ECE_{TL0-1}	ECE_{TL0-2}	$ECE_{TL=1}$	TL_0	ECE_{TL0-1}	ECE_{TL0-2}	$ECE_{TL=1}$
2008	3.335	0.4977	0.6539	0.7062	3.310	0.5016	0.6578	0.7062
2009	3.305	0.5024	0.6586	0.7062	3.280	0.5063	0.6625	0.7062
2010	3.275	0.5071	0.6633	0.7062	3.250	0.5110	0.6672	0.7062
2011	3.245	0.5118	0.6679	0.7062	3.220	0.5157	0.6718	0.7062
2012	3.215	0.5165	0.6726	0.7062	3.190	0.5204	0.6765	0.7062
2013	3.185	0.5212	0.6773	0.7062	3.160	0.5251	0.6812	0.7062
2014	3.155	0.5258	0.6820	0.7062	3.130	0.5298	0.6859	0.7062
2015	3.125	0.5305	0.6867	0.7062	3.100	0.5344	0.6906	0.7062

Based on these calculations, the annual carbon sequestration of each coastal area was calculated, and the results were summed to obtain the total carbon sequestration of each coastal area between 2008 and 2015. The values are shown in Table 4.10. The total carbon sequestration of 11 coastal areas in China during the study period ranged from 1520.38 to 1571.90 10^5 tC, and the annual average carbon sequestration was 1544.91 10^5 tC. According to the Kyoto Protocol of the United Nations Framework Convention on Climate Change, the expenditure of industrialized countries to reduce CO_2 is estimated to be USD150−600 per ton, and the corresponding value of carbon sequestration per year can be calculated. Table 4.10 shows that the average value of the annual reduction in emissions in 11 coastal areas in China was USD23,174−92,694 million.

Fig. 4.4 shows the evolution of marine carbon sinks in China's coastal areas between 2008 and 2015 based on the data in Table 4.9. Zhejiang

Table 4.7 Carbon contents of four common algae (Harrison, Druehl, Lloyd, & Thompson, 1986; Lapointe & Littler, 1992; Zhou, Yang, & Liu, 2002).

Category	Carbon content (%)
Kelp	31.17
Undaria pinnatifida	30.70
Porphyra	27.39
Gracilaria	28.40

Table 4.8 Dry shell weight coefficients and carbon contents of four common shells (Zhou, Yang, & Liu, 2002).

Category	Dry shell weight coefficient (%)	Carbon content (%)
Scallop	56.60	11.44
Oysters	63.81	11.52
Clam	45.01	11.40
Mussel	70.64	12.68

province ranked first in terms of total carbon sink, followed by Fujian, Shandong, Guangdong, and other provinces. The evolution curve of the marine carbon sink in Zhejiang province was unusual, and increased annually from 2008 to 2012, with the greatest increase in 2009, and the peak value of 417.31 10^5 tC obtained in 2012; it decreased from 2012 to 2015 and reached 409.54 10^5 tC. Generally, the overall increase was greater than the decline. The carbon sink of Hainan province declined slightly in 2009, increased annually from 2010 to 2012, reached 141.47 10^5 tC in 2012, decreased to 139.10 in 2013, and increased again in 2014 and 2015, peaking at 160.60 10^5 tC in 2015. The carbon sinks of Tianjin and Shanghai municipalities were relatively small; the evolutionary trends were not clear from Fig. 4.4 but can be observed in Table 4.9. The carbon sink of Tianjin tended to decrease from 2.57 in 2008 to 2.12 in 2012, although a small increase was observed in 2011; this value increased sharply to 6.76 in 2013, decreased to 5.68 in 2014, and increased slightly to 5.78 in 2015. The carbon sink of Shanghai decreased annually, except in 2009 and 2012; it declined from 2.96 in 2008 to 2.07 in 2015. Carbon sinks of other areas followed a declining trend. Considering 2008 as the baseline period, changes in carbon sinks in coastal areas during the study period were investigated (Table 4.11). Except for the increases in Tianjin, Zhejiang, and Hainan, the carbon sinks in other areas decreased during the study period. Zhejiang province presented the largest increase in carbon sink at 64.03 10^5 tC, while the greatest reduction in carbon sink occurred in Guangdong province, 29.94 10^5 tC.

Table 4.9 Estimates of marine carbon sinks of 11 coastal areas in China from 2008 to 2015 (unit: 10^5 tC).

	Hebei	Liaoning	Tianjin	Shandong	Jiangsu	Shanghai	Zhejiang	Fujian	Guangxi	Guangdong	Hainan
2008	34.83	144.91	2.57	223.25	80.16	2.96	345.50	271.70	94.92	209.30	133.40
2009	34.29	138.04	2.22	219.19	78.69	3.13	382.45	268.36	93.35	198.20	133.09
2010	33.78	138.27	2.09	215.24	78.67	3.00	393.66	268.06	90.85	195.53	133.96
2011	33.19	143.64	2.23	215.47	81.58	2.91	411.33	261.92	88.72	193.27	137.64
2012	32.70	144.34	2.12	210.94	76.20	3.18	417.31	256.34	86.52	195.65	141.47
2013	29.45	139.20	6.76	198.79	73.63	2.52	410.13	250.88	82.21	187.69	139.10
2014	30.16	137.63	5.68	200.05	71.92	2.53	411.84	252.91	81.27	185.83	150.26
2015	31.03	142.11	5.78	196.28	71.81	2.07	409.54	245.84	78.06	179.35	160.60

Table 4.10 Carbon sequestration and value of 11 coastal areas in China from 2008 to 2015.

Year	2008	2009	2010	2011	2012	2013	2014	2015
Carbon sequestrations/10^5 t	1543.51	1551.01	1553.12	1571.90	1566.78	1520.38	1530.09	1522.47
Value/USD100 million	231.53—	232.65—	232.97—	235.79—	235.02—	228.06—	229.51—	228.37—
	926.10	930.61	931.87	943.14	940.07	912.23	918.05	913.48

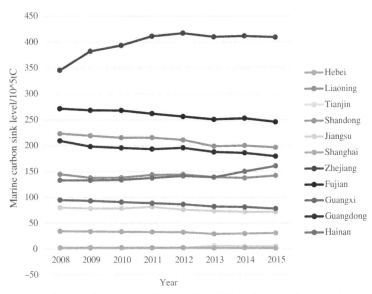

Figure 4.4 Evolution of marine carbon sinks of China's coastal areas during 2008 and 2015.

The marine carbon sinks of coastal areas can be divided into six levels: <10, $10-50$, $50-100$, $100-200$, $200-300$, and >300 10^5 tC. Regional distribution maps showing carbon sink levels between 2008 and 2015 were drawn by ARCGIS software (Fig. 4.5). Except for Shandong and Guangdong provinces, where the carbon sinks decreased from $200-300$ in 2008 to $100-200$ 10^5 tC in 2015, the carbon sink levels in the other coastal areas remained constant. Overall, high carbon sink levels (>200 10^5 tC) were distributed in Zhejiang and Fujian, while low carbon sink levels (<50 10^5 tC) were distributed in Hebei, Tianjin, and Shanghai. The carbon sinks of Jiangsu and Guangxi were $50-100$ 10^5 tC and those of Liaoning, Guangdong, and Hainan were $100-200$ 10^5 tC.

Tables 4.12 and 4.13 show the total amount and composition of marine carbon sinks in China's coastal areas in 2008 and 2015, respectively. The percentages represent the proportions of each carbon sink component to the total carbon sink. The data in Tables 4.12 and 4.13 indicate that phytoplankton carbon sequestration accounts for the majority of marine carbon sinks, with a proportion exceeding 98%. This shows that the primary productivity of phytoplankton plays an important role in the marine carbon sink. The remaining 2% is shared by algae carbon sequestration and shellfish carbon sequestration. Algae and shellfish in

Table 4.11 Changes in ocean carbon sequestration in China's coastal provinces and cities during the study period.

	Hebei	Liaoning	Tianjin	Shandong	Jiangsu	Shanghai	Zhejiang	Fujian	Guangxi	Guangdong	Hainan
Change/10^5 tC	− 3.80	− 2.80	+ 3.21	− 26.98	− 8.35	− 0.89	+ 64.03	− 25.87	− 16.86	− 29.94	+ 27.20

Source: Based on data for 2008.

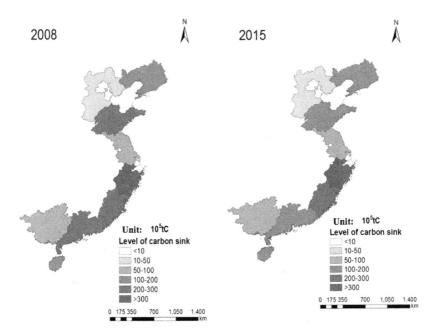

Figure 4.5 Regional distribution of marine carbon sinks in 11 coastal areas in China from 2008 to 2015.

Shandong and Fujian provinces contributed most to total carbon sequestration. The total algae carbon sequestration exceeded 0.4×10^5 tC in these two provinces, while that of shellfish exceeded 1.0×10^5 tC. Based on carbon sequestration in the total carbon sink, algae carbon sequestration in each area accounted for less than 0.3% of the total carbon sinks of all areas, while shell carbon sequestration accounted for a relatively high proportion, exceeding 0.7% or even 1% in some areas.

4.5.3.2 Theil index and its decomposition

Theil index or Theil's entropy measure is a way to examine inequality based on the concept of entropy in information theory (Zhang, 2004). The bigger the Theil index, the more significant the regional difference. One advantage of using the Theil index to measure inequality differences is that it can measure the contribution of intraregional and interregional differences to the total difference. In this section, the Theil index was used to analyze the regional differences of marine carbon sinks in 11 coastal areas in China.

Table 4.12 Total marine carbon sinks and compositions in China's coastal areas in 2008.

2008	Total carbon sink (10^5 tC)	Phytoplankton carbon sequestration		Algae carbon sequestration		Shellfish shells carbon sequestration	
		Total (10^5 tC)	Percentage	Total (10^5 tC)	Percentage	Total (10^5 tC)	Percentage
Hebei	34.83	34.65	99.49	0	0	0.18	0.51
Liaoning	144.91	143.65	99.13	0.20	0.14	1.06	0.73
Tianjin	2.57	2.57	99.88	0	0	0.003	0.12
Shandong	223.25	220.96	98.97	0.41	0.18	1.88	0.84
Jiangsu	80.16	79.78	99.53	0.019	0.02	0.36	0.45
Shanghai	2.96	2.96	100.00	0	0	0	0
Zhejiang	345.50	345.00	99.85	0.026	0.01	0.48	0.14
Fujian	271.70	269.81	99.30	0.41	0.15	1.48	0.54
Guangxi	94.92	94.48	99.54	0	0	0.44	0.46
Guangdong	209.30	208.09	99.42	0.045	0.02	1.16	0.55
Hainan	133.40	133.35	99.96	0.024	0.02	0.023	0.02

Table 4.13 Total marine carbon sinks and compositions in China's coastal areas in 2015.

2008	Total carbon sink (10^5 tC)	Phytoplankton carbon sequestration		Algae carbon sequestration		Shellfish shell carbon sequestration	
		Total (10^5 tC)	Percentage	Total (10^5 tC)	Percentage	Total (10^5 tC)	Percentage
Hebei	31.03	30.72	99.00	0	0	0.31	1.00
Liaoning	142.11	140.61	98.94	0.28	0.20	1.21	0.85
Tianjin	5.78	5.78	100.00	0	0	0.0017	0.03
Shandong	196.28	193.07	98.36	0.54	0.28	2.66	1.36
Jiangsu	71.81	71.35	99.36	0.02	0.03	0.44	0.61
Shanghai	2.07	2.07	100.00	0	0	0	0
Zhejiang	409.54	408.94	99.85	0.04	0.01	0.55	0.13
Fujian	245.84	243.25	98.95	0.72	0.29	1.87	0.76
Guangxi	78.06	77.46	99.23	0	0	0.60	0.77
Guangdong	179.35	177.88	99.18	0.06	0.03	1.41	0.79
Hainan	160.60	160.52	99.95	0.03	0.02	0.05	0.03

First, 11 coastal areas in China were divided into three regions: the northern coastal region, the eastern coastal region, and the southern coastal region. The northern coastal region is composed of Hebei, Liaoning, Tianjin, and Shandong; the eastern coastal region is composed of Jiangsu, Shanghai, and Zhejiang; and the southern region consists of Fujian, Guangxi, Guangdong, and Hainan. The Theil index formula is as follows:

$$T = \sum_{j}\sum_{j}\frac{I_{ij}}{I} \times \ln\left(\frac{I_{ij}/I}{N_{ij}/N}\right) \tag{4.21}$$

where T is the Theil index of 11 coastal areas overall; I represents the marine carbon sink level; N refers to the gross ocean product of each coastal area; the subscript i represents the division of the three major regions; the subscript j refers to each area in each region; I_{ij} is the marine carbon sink level of provincial administrative district j in region i, and N_{ij} is the gross ocean product of coastal area j in region i.

The Theil index, T, is the sum of the intraregional difference, T_w, and the interregional difference, T_b. The formula is as follows:

$$T = T_b + T_w \tag{4.22}$$

$$T_b = \sum_{i}\frac{I_i}{I} \times \ln\left(\frac{I_i/I}{N_i/N}\right) \tag{4.23}$$

$$T_w = \sum_{i}\frac{I_i}{I}\left[\sum_{j}\frac{I_{ij}}{I_i}\ln\left(\frac{I_{ij}/I_i}{1/N_i}\right)\right] = \left(\sum_{i}\frac{I_i}{I}\right) \times T_{wi} \tag{4.24}$$

where T_b is the overall interregional difference, T_w is the overall intraregional difference, I_i is the marine carbon sink level in region i, N_i is the gross ocean product of each coastal area in region i, and T_{wi} is the intraregional difference in region i.

Based on calculations of the overall Theil index and its decomposition terms, the contribution of interregional and intraregional differences to the overall difference can be determined. The contribution of the interregional marine carbon sink level to the overall difference in marine carbon sink levels among coastal areas in China is as follows:

$$D_b = \frac{T_b}{T} \tag{4.25}$$

The contribution of the difference in marine carbon sink level within region i to the overall difference in marine carbon sink level in coastal areas of China is as follows:

$$D_i = \frac{I_i}{I} \times \frac{T_i}{T} \qquad (4.26)$$

Accordingly, overall differences in marine carbon sinks of 11 coastal areas and decomposed intraregional and interregional differences between 2008 and 2015 were calculated and the intraregional difference within each region was obtained. Curves showing changes in the Theil index are shown in Figs. 4.6 and 4.7.

Fig. 4.6 shows variation in the Theil index for the overall, interregional, and intraregional differences in marine carbon sinks in China's coastal areas from 2008 to 2015. The overall difference fluctuated during the research period, while there was little difference in the Theil index between 2015 and 2014. In terms of the magnitude of increase and decrease, the biggest decrease was 0.0386 in 2009 compared with the previous year, while the largest increase was 0.0394 in 2010. Taking 2008 as the baseline period, the overall Theil index increased by 0.0175 during the research period, indicating increasing interregional differences. By breaking down the overall difference into intraregional and interregional differences, the Theil index of intraregional differences followed a similar trend to the overall difference. Taking 2008 as the baseline period, the Theil index increased by 0.0384, indicating that the overall regional difference in coastal areas increased, and the proportion of intraregional difference in the overall difference was larger, meaning that the variation of the overall difference was mainly caused by the intraregional differences in the three coastal regions. Regarding interregional differences, the Theil index tended to decrease during the research period; taking 2008 as the baseline period, it decreased by 0.0210, indicating that differences in marine carbon sinks among the three regions decreased.

Fig. 4.7 shows the differences in marine carbon sinks among the northern, eastern, and southern coastal regions in China from 2008 to 2015. The Theil index of differences in the eastern region followed a downward trend, indicating that differences among areas in this region decreased during the research period. Differences among areas in the northern and southern regions followed an upward trend, indicating that differences among areas in these regions increased during the research period. Comparison of the three curves revealed that the annual average Theil index was highest in the eastern region, followed by the southern and northern regions. This shows that

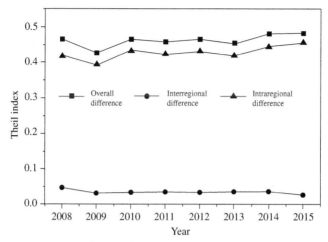

Figure 4.6 Overall, interregional, and intraregional differences in marine carbon sinks in China's coastal areas from 2008 to 2015.

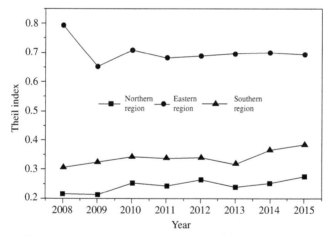

Figure 4.7 Differences in marine carbon sinks among northern, eastern, and southern coastal regions in China between 2008 and 2015.

differences among areas in the eastern region were the largest, followed by the southern and northern regions successively. For the eastern region, the Theil index decreased from 0.7939 in 2008 to 0.6938 in 2015; the value of 0.6538 in 2009 was the lowest, indicating that the interregional difference in the eastern region was the smallest in 2009, and the decrease in the Theil index in 2009 was the largest, at 0.1400. The Theil index in the southern region increased from 0.3067 to 0.3855, with 0.3187 in 2013 being the lowest value. In the northern region, the Theil index increased by 0.0606, from 0.2153 to 0.2759.

Table 4.14 Contribution of intraregional and interregional differences in the overall Theil index of China's coastal areas from 2008 to 2015.

Year	Overall difference	Contribution rate (%)			
		Interregional	Northern	Eastern	Southern
2008	0.4651	10.14	12.16	47.40	30.30
2009	0.4265	7.39	12.62	45.89	34.10
2010	0.4658	7.24	13.58	46.56	32.61
2011	0.4581	7.64	13.28	47.02	32.06
2012	0.4652	7.24	14.09	46.96	31.71
2013	0.4547	7.85	12.83	48.89	30.42
2014	0.4803	7.50	12.77	46.35	33.39
2015	0.4826	5.43	14.09	45.65	34.84

Contribution of intraregional and interregional differences to the overall total Theil index was calculated according to formulas (4.21−4.24) and the results are shown in Table 4.14. The contribution of interregional differences was relatively small and tended to decrease during the research period, with an average value of only 7.55%. The remaining proportion was due to the intraregional differences. Within the three regions, the contribution of the intraregional difference in the eastern region was the largest, with an annual average of 46.84%, followed by the southern region, with an annual average of 32.43%. The contribution in the northern region was the smallest, with an annual average of only 13.18%. Therefore the intraregional differences within eastern and southern regions had significant impacts on the overall difference. Taking 2008 as the baseline period, the contribution of the difference within the northern region increased by 1.93%, that within the eastern region decreased by 1.75%, and that within the southern region increased by 4.54%.

4.6 Conclusion and suggestions

4.6.1 Conclusion

Estimation of the marine carbon sinks of 11 coastal areas in China from 2008 to 2015, and analysis of regional differences and evolution trends using the Theil index and its decomposition led to following conclusions:

1. Regarding the total marine carbon sink, the total carbon sequestration of 11 coastal areas in China during the research period ranged from

1520.38 to 1571.90 10^5 tC, with an average annual carbon sequestration of 1544.91 10^5 tC. The average annual reduced emission had a value of USD23,174−92,694 billion. The high values for marine carbon sinks were distributed in Zhejiang and Fujian, while the low values were in Hebei, Tianjin, and Shanghai. During the research period, the carbon sinks of Hainan, Tianjin, and Zhejiang tended to grow, while those of other coastal areas tended to decrease. The carbon sink level of Zhejiang province ranked highest, with an annual average of 397.72 10^5 tC, with an increase of 18.53% during the research period. The carbon sink was divided into six levels: <10, 10−50, 50−100, 100−200, 200−300, and >300 10^5 tC. During the research period, except for Shandong and Guangdong, the carbon sink levels of other provinces did not change.

2. Regarding the composition of marine carbon sinks, the primary productivity of phytoplankton accounted for about 99%, and the remaining 1% involved carbon sequestration from algae and shellfish. The primary productivity of phytoplankton is of great significance to the maintenance of marine carbon sinks. The main factors affecting photosynthesis in phytoplankton are light, temperature, and nutrients. With global warming, increasing ocean temperatures will lead to a decrease in phytoplankton and carbon sinks, indirectly promoting climate warming. Considering the decrease in marine carbon sinks in most coastal areas in China, local governments must recognize the importance of conserving energy and reducing emissions, organically combine economic growth with sustainable development, optimize the energy utilization structure, contribute to preventing global warming, and work together to build a beautiful China with green mountains and green waters. Furthermore, optimizing the structure of shellfish and algae aquaculture and vigorously developing the aquaculture industry with high carbon sink capacity may also help to slow the greenhouse effect.

3. China's coastal provinces and cities are divided into three regions: the northern coastal region, the eastern coastal region, and the southern coastal region. By calculating the Theil index of marine carbon sinks in China's coastal areas from 2008 to 2015, taking 2008 as the baseline, the overall Theil index increased by 0.0175, indicating that interregional differences in marine carbon sinks increased during the research period. By breaking down the Theil index into intra- and interregional differences, we found that the intraregional difference

increased by 0.0384, while the interregional difference decreased by 0.0210 during the research period. The contribution of intraregional differences to the overall difference was about 93%, while that of interregional differences to the total difference was only about 7%. This showed that the change in the overall difference of marine carbon sinks in China's coastal areas is caused by intraregional differences in the three major coastal regions.

4. Regarding differences within each region, that in the eastern region decreased during the research period, while those in the northern and southern regions increased. Regarding the contribution of intraregional differences within the three regions, that of the eastern region was the largest, with an average annual value of 46.84%, followed by the southern region, with an average annual value of 32.43%, and the northern region, with an average annual value of only 13.18%. Thus intraregional differences in the eastern and northern regions have significant impacts on the overall difference. Taking 2008 as the baseline period, the contribution of intraregional differences in the northern and southern regions increased by 1.93% and 4.54%, respectively, while that in the eastern region decreased by 1.75%.

4.6.2 Suggestions

1. Publicity should be enhanced to increase awareness of marine carbon sinks. As they represent a new concept, time is needed for marine carbon sinks to be recognized and accepted by society. Coastal fishermen tend to pursue fishery cultivation and catches unilaterally in fishery production activities. It is accepted that the high productivity of fisheries increases the carbon sequestration of the marine carbon sink. However, in contrast, the practice of increasing fishery production will break the lead to imbalances in marine ecology and seriously damage the natural ecological environment of the local sea area. Therefore efforts must be strengthened to enhance the social recognition of marine carbon sinks. Leaders at all levels and local governments should recognize the necessity and urgency of developing marine carbon sinks and actively participate in their development. Through the joint efforts of governments and the population, marine carbon sinks will boost and transform low-carbon economies.

2. Accounting standards and evaluation systems of marine carbon sink should be established and optimized, such as formulating accounting

standards for physical carbon sequestration in seawater, accounting standards for carbon sequestration in marine organisms, and improving the audit of carbon emission reduction indexes. Marine carbon sinks represent a new research field that is still in its infancy. There are no unified international norms and standards, or mature experience for reference. Therefore there is an urgent need to establish and improve their measurement and evaluation. Compared with forest carbon sinks, marine carbon sinks possess regional characteristics due to their huge volume, and many factors may affect marine carbon sinks. Hence, it is difficult to establish a unified accounting and evaluation standard for marine carbon sinks. In view of this, local governments in coastal areas can establish standards and evaluation systems for marine carbon sinks in line with the actual situation of the sea areas under their jurisdiction.

3. A marine carbon sink trading system should be established and improved. To meet energy-saving requirements, reduced emissions, and sustainable development, the management of marine resources and environments should be strengthened, legal mechanisms of the carbon sink market should be established and improved, and the mode of marine carbon sink trading should be actively explored and innovated in line with China's national conditions and international trading rules. Pilot projects can be established in some areas first and subsequently promoted in the whole country at the right time. Meanwhile, businesses and the public can be encouraged to actively participate in the construction of blue ocean carbon sinks.

4. The technological innovation and development of marine carbon sinks should be promoted. Abundant renewable energy reserves exist in the ocean, which will comprise an important research field in the future low-carbon social construction. To meet the growing energy demands in China, the government should encourage and support research into technology for the development of renewable marine energy, especially strengthening the development of marine renewable energies such as tidal energy, wave energy, salinity energy, sea temperature energy, and sea wind energy to promote the adjustment of China's energy structure and improve the carbon sink capacity of the ocean.

References

Alpert, S. B., Spencer, D. F., & Hidy, G. (1992). Biospheric options for mitigating atmospheric carbon dioxide levels. *Energy Conversion and Management, 33*(5−8), 729−736.

Bacher, C., Bioteau, H., & Chapelie, A. (1995). Modelling the impact of a cultivated oyster population on the nitrogen dynamics: The Thau Lagoon case (France). *Ophelia*, *42*, 29−54.

Bakker, D. C. E., de Baar, H. J. W., & de Jong, E. (1999). The dependence on temperature and salinity of dissolved inorganic carbon in East Atlantic surface waters. *Marine Chemistry*, *65*, 263−280.

Cauwet, G. (1977). Organic chemistry of seawater particulates-concepts and developments. *Marine Chemistry*, *5*(4−6), 551−552.

Cerrato, R. M., Caron, D. A., Lonsdale, D. J., Rose, J. M., & Schaffner, R. A. (2004). Effect of the northern quahog *Mercenaria mercenaria* on the development of blooms of the brown tide alga *Aureococcus anophagefferens*. *Marine Ecology Progress Series*, *281*, 93−108.

Chao, M., Quan, W. M., Li, C. H., & Chen, Y. H. (2005). Changes in trophic level of marine catches in the East China Sea region. *Marine Science*, *29*(9), 51−55.

Chapman, A. R. O. (1974). Ecology of macroscopic marine algae. *Annual Review of Ecology & Systematics*, *5*, 65−80.

Chisholm, S. W. (2000). Oceanography-Stirring times in the Southern Ocean. *Nature*, *407*(6805), 685−687.

Clarke, G. L., Ewing, G. C., & Lorenzen, C. J. (1970). Spectra of backscattered light from the sea obtained from aircraft as a measure of chlorophyll concentration. *Science*, *167* (3921), 1119−1121.

Committee on Global Change. (1988). *Toward an understanding of global change* (p. 56) Washington, DC: National Academy Press.

Eatherall, A., Naden, P. S., & Cooper, D. M. (1998). Simulating carbon flux to the estuary: The first step. *The Science of the Total Environment*, *210/211*, 519−533.

Falkowski, P. G., Barber, R. T., & Smetacek, V. V. (1998). Biogeochemical controls and feedbacks on ocean primary production. *Science*, *281*(5374), 200−207.

Gong, H. D., Zhang, Z. B., & Zhang, C. (2006). Multilayer study of carbon dioxide system in the surface waters of the Yellow Sea in spring. *Chinese Journal of Oceanology and Limnology*, *25*(1), 1−15.

González, J. M., Fernández-Gómez, B., Fernàndez-Guerra, A., Gómez-Consarnau, L., Sánchez, O., Coll-Lladó, M., . . . Pedrós-Alió, C. (2008). Genome analysis of the proteorhodopsin-containing marine bacterium *Polaribacter* sp. MED152 (Flavobacteria): A tale of two environments. *Proceedings of the Natural Academy of Science USA*, *105*, 8724−8729.

Harrison, P. J., Druehl, L. D., Lloyd, K. E., & Thompson, P. A. (1986). Nitrogen uptake kinetics in three year-classes of *Laminaria groenlandica* (*Laminariales phaeophyta*). *Marine Biology*, *93*(1), 29−35.

Hyde, K. J. M., O'Reilly, J. E., & Oviatt, C. A. (2008). Evaluation and application of satellite primary production models in Massachusetts Bay. *Continental Shelf Research*, *28* (10−11), 1340−1351.

IPCC. (2007). Climate Change 2007: The physical science basis. In: *Contribution of working group I to the fourth assessment report of the intergovernmental panel on climate change*. UK: Cambridge University Press.

Keeling, C. D., Baestow, R. B., Carter, A. F., Piper, S. C., Whorf, T. P., Heimann, M., . . . Roeloffzen, H. (1989). A three-dimensional model of CO_2 transport based on observed winds. I. Analysis of observational data. *American Geophysical Union, Monograph*, *55*, 165−234.

Khatiwala, S., Primeau, F., & Hall, T. (2009). Reconstruction of the history of anthropogenic CO_2 concentrations in the ocean. *Nature*, *462*, 346−349.

Lapointe, B. E., & Littler, D. S. (1992). Nutrient availability to marine macro algae in siliciclastic versus carbonate-rich coastal waters. *Estuaries & Coasts*, *15*(1), 75−82.

Lee, K., & Kenneth, H. (1997). Effects of in situ light reduction on the maintenance, growth and partitioning of carbon resources in *Thalassia testudinum* Banks ex Konig. *Journal of Experimental Marine Biology and Ecology, 210*(1), 53–73.

Leng, Y. Z., He, L. T., & Wei, Q. H. (1984). Population biology and resource estimation of copperfish in the upper reaches of the Yangtze River after the interception of the Gezhouba Water Control Project. *Freshwater Fisheries, 5*, 21–25.

Li, L., Hao, T., & Chi, T. (2017). Evaluation on China's forestry resources efficiency based on big data. *Journal of Cleaner Production, 142*, 513–523.

Liu, H. Y., Ren, G. R., & Zhang, S. L. (2011). Discussion on the contribution and development of carbon sink fisheries to marine low carbon economy. In: *The 5th China animal husbandry science and technology forum*, Chongqing, China.

Liu, X., Mao, G. Z., Ren, J., & Li, R. Y. M. (2014). How might China achieve its 2020 emissions target? A scenario analysis of energy consumption and CO_2 emissions using the system dynamics model. *Journal of Cleaner Production, 103*, 401–410.

Lv, R. H., Xia, B., Li, B. H., & Fei, Z. L. (1999). The fluctuations of primary productivity in Bohai sea waters over ten years. *Advances in Marine Science, 17*(3), 80–86.

Macroy, C., & McMillan, C. (1977). Production of ecology and physiology of seagrasses. *Marine Science, 4*, 53–81.

Mann, K., & Chapman, A. (1975). Primary production of marine macrophytes. *International biology Progress, 3*, 1–15.

Mikaloff-Fletcher, S. E., Gruber, N., Jacobson, A. R., & Doney, S. C. (2006). Inverse estimates of anthropogenic CO_2 uptake, transport, and storage by the ocean. *Global Biogeochemistry Cycles, 20*(2), GB2002.

Nielsen, E. S. (1952). The use of radio-active carbon (C^{14}) for measuring organic production in the sea. *ICES Journal of Marine Science, 18*(2), 117–140.

Pauly, D., Christensen, V., Dalsgaard, J., Froese, R., Jr, & Francisco, T. (1998). Fishing down marine food webs. *Science, 279*(5352), 860–863.

Peterson, B. J., & Heck, K. L. (1999). The potential for suspension feeding bivalves to increase seagrass productivity. *Journal of Experimental Marine Biology and Ecology, 240*(1), 37–52.

Quay, P. D., Sonnerup, R., Westby, T., Stutsman, J., & McNichol, A. (2003). Changes in the $^{13}C/^{12}C$ of dissolved inorganic carbon in the ocean as a tracer of anthropogenic CO_2 uptake. *Global Biogeochemical Cycles, 17*(1), 4-1–4-20.

Riisgard, H. (1991). Filtration rate and growth in the blue mussel *Mytilus edulis* Linnaeus 1758 dependence on algal concentration. *Journal of Shellfish Research, 10*(1), 29–36.

Seeyave, S., Lucas, M. I., Moore, C. M., & Poulton, A. J. (2007). Phytoplankton productivity and community structure in the vicinity of the Crozet Plateau during austral summer 2004/2005. *Deep Sea Research Part II: Topical Studies in Oceanography, 54*, 2020–2044.

Takashi, Y., Ichiro, T., & Ken, F. (2001). Active erosion of *Undaria pinnatifida* Suringar (Laminariales, Phaeophyceae) mass cultured in Otsuchi Bay in northeastern Japan. *Journal of Experimental Marine Biology and Ecology, 266*, 51–65.

Tang, Q. S., Guo, X. W., Sun, Y., & Zhang, B. (2007). Ecological conversion efficiency and its influencers in twelve species of fish in the Yellow Sea Ecosystem. *Journal of Marine Systems, 67*(3–4), 282–291.

Teira, E., Mouriño, B., Marañón, E., Pérez, V., Pazó, M. J., Serret, P., ... Fernández, E. (2005). Variability of chlorophyll and primary production in the Eastern North Atlantic Subtropical Gyre: Potential factors affecting phytoplankton activity. *Deep Sea Research, Part I: Oceanographic Research Papers, 52*(4), 569–588.

Tilstone, G. H., Taylor, B. H., Blondeau-Patissier, D., Powell, T., Groom, S. B., Rees, A. P., & Lucas, M. L. (2015). Comparison of new and primary production models using SeaWiFS data in contrasting hydrographic zones of the northern North Atlantic. *Remote Sensing of Environment, 156,* 473−489.

Tin, H. C., Lomas, M. W., & Ishizaka, J. (2015). Satellite-derived estimates of primary production during the Sargasso Sea winter/spring bloom: Integration of in-situ time-series data and ocean color remote sensing observations. *Regional Studies in Marine Science, 3,* 131−143.

Vitousek, P. M., Mooney, H. A., Lubchenco, J., & Melillo, J. M. (1997). Human domination of Earth's ecosystems. *Science, 277*(5325), 494−499.

Watson, R. H., Rodhe, H., & Oesehager, H. (1990). *Greenhouse gases and aerosols* (pp. 1−40). New York: Cambridge University Press.

Whittaker, R. H. (1975). *Communities and ecosystems.* New York: Macmillan Publishing Co.

Wong, W. H., Levinton, J. S., Twining, B. S., Fisher, N. S., Kelaher, B. P., & Alt, A. K. (2003). Assimilation of carbon from a rotifer by the mussels *Mytilus edulis* and *Perna viridis*: A potential food web link. *Marine Ecology Progress Series, 253,* 175−182.

Wu, J. M., Lou, B. Y., Zhao, H. T., Li, L., Chen, Y. X., & Wang, J. W. (2011). Preliminary assessment of fish stock in the Chishui River. *Journal of Hydroecology, 32* (3), 99−103.

Yang, H. S., Yuan, X. T., Zhou, Y., Mao, Y. Z., Zhang, T., & Liu, Y. (2005). Effects of body size and water temperature on food consumption and growth in the sea cucumber *Apositchopus japonicus* (Selenka) with special reference to aestivation. *Aquaculture Research, 36,* 1085−1092.

Yang, Y. L., Li, Y., & Pan, J. (2004). Characteristics of an open complex giant system— Carbon cycling system in the ocean. *Complex Systems and Complexity Science, 01,* 68−77.

Ye, C. C., & Tang, Q. S. (1980). Estimation of the population of the Pacific Yellow Sea population. *Journal of Zoology, 02,* 3−7.

Zeebe, R. E., & Wolf-Gladrow, D. (2001). *CO_2 in seawater: equilibrium, kinetics, isotopes* (Vol. 65, p. 346). Netherlands, Amsterdam: Elsevier Oceanography Book Series.

Zhang, B., Tang, Q., & Jin, X. (2007). Decadal-scale variations of trophic levels at high trophic levels in the Yellow Sea and the Bohai Sea ecosystem. *Journal of Marine Systems, 67*(3-4), 304−311.

Zhang, H. M. (2004). Empirical analysis of regional difference of rural resident's agricultural tax. *Journal of China Agricultural University (Social Science Edition), 01,* 19−23.

Zhang, N., & Choi, Y. (2013). Total-factor carbon emission performance of fossil fuel power plants in china: A metafrontier non-radial Malmquist index analysis. *Energy Economics, 40*(2), 549−559.

Zhang, Y. R., Ding, Y. P., Li, T. J., Xue, B., & Guo, Y. M. (2016). Annual variations of chlorophyll A and primary productivity in the East China Sea. *Oceanologia Et Limnologia Sinica, 47*(01), 261−268.

Zhou, Y., Yang, H. S., & Liu, S. L. (2002). Chemical composition and net organic production of cultivated and fouling organisms in Sishili Bay and their ecological effects. *Journal of Fisheries of China, 26*(1), 21−27.

CHAPTER FIVE

Comprehensive benefit evaluation of marine resource utilization in China

5.1 Introduction

With the shortage of land resources, rapid population growth, eco-logical damage, and environmental deterioration becoming increasingly serious issues, the intensity of marine development is increasing (Alfonso et al., 2001; Cavanagh et al., 2016; Cognetti & Maltagliati, 2010; Francisco, Rosa, Francisco, & Carlos, 2014). Further, there is a growing importance of the marine economy in the development of national econ-omy and society. China is also becoming increasingly aware of the impor-tance of developing its marine economy. To clear the bottleneck of resources and environment and explore new ways of transformation and development, the Chinese government has set up a special chapter in the 13th Five-Year Plan to promote marine economic development. The aim is to optimize the spatial distribution of the marine economy and make it a new engine for the transformation and upgradation of China's economy, especially in the eastern coastal region. The oceans are an important part of the global life support system. Marine resources are a treasure house of resources because of their large volume, variety, and distribution. The oceans are an important base for providing strategic reserve resources for national economic and social development and play a strong supporting role to marine resources, such as food, water, minerals, and energy needed for human survival and development (Giselle et al., 2007; Paul & Teresa, 2003; Pontecorvo & Wilkinxon, 1980). In recent years, 11 areas along the coast of China have gradually increased their efforts to develop marine resources. This has brought economic and social benefits to coastal areas, besides causing negative effects on marine ecological environment. The latter has resulted in a series of problems, such as disorderly exploitation of resources, repeated exploitation, serious environmental pollution, blind

Sustainable Marine Resource Utilization in China.
DOI: https://doi.org/10.1016/B978-0-12-819911-4.00005-9

reclamation of land from the sea, and low efficiency of development and utilization (Costanza, 1999; Fathab, 2015; Porobic et al., 2017). The exploitation and utilization of marine resources are related to all the aspects of the national economy. It has not only economic benefits but also social, ecological, and environmental ones. To realize the sustainable development of a marine economy, it is necessary to improve the comprehensive economic—social—environmental benefits from the utilization of marine resources. This research intends to build a comprehensive benefit evaluation model of marine resource development and utilization in coastal areas; quantitatively explore the comprehensive benefits; and, more pertinently, enhance the comprehensive benefits of resource utilization.

5.2 Literature review

The comprehensive benefits of marine resource development and utilization include three aspects: economic, social, and environmental benefits. Duan and Xu (2009) constructed a comprehensive benefit evaluation system of marine resource utilization from the perspective of industrial benefits. Based on the niche theory and polygon synthesis index method, they analyzed and evaluated the comprehensive benefit level and regional differences of marine resource utilization in their researched area. Gao et al. (2017) considered four main types of marine development in Jiangsu province as examples to explore a more reasonable, objective, and suitable marine benefit evaluation system. The total benefit evaluation includes social and economic benefits, as well as the value of ecosystem services, ecosystem loss, and ecological benefits. The comprehensive benefit index system is constructed on the basis of social, economic, resource, and environmental benefits. Then, the comprehensive index method and gray correlation method are used to evaluate the comprehensive benefits of different types of marine development; this provides a quantitative reference point for the government to formulate relevant policies of marine environmental protection in the Bohai Sea. Wei, Li, and Zhao (2012) studied and constructed the comprehensive benefit evaluation index system of China's marine resource development and utilization; further, they evaluated quantitatively the comprehensive benefit status of China's marine resource development and utilization by using the entire-array-polygon index evaluation method. The results show that the composite index was

0.381 in 2006 and 0.508 in 2009. Daniel, Andrew, Ian, and Jin (2014) introduced a method to analyze the cost of operating marine systems under different conditions. Data obtained from a previous Monte Carlo analysis are used to assess the operational costs of various maintenance and inspection policies. The aim was to show that the Monte Carlo analysis could be adapted to provide information on factors affecting operational costs for decision-making to optimize the efficiency of ocean systems. Scholars in this field have measured the values of direct and indirect utilization of marine biological resources. Fu, Wang, and Jiang (2018) evaluated the value of marine living resources in Shandong province in 2014 by using the market valuation method, coefficient method, changes in productivity approach, result reference method, and conditional value method. In 2014 the total economic value of marine living resources in Shandong province was USD 470,533 million; the direct use value was USD 9387.2 million; the indirect use value was USD 30,347 million; and the social value was USD 7319 million. Indirect use value accounted for the largest proportion (64.49%), while direct use value accounted for 19.95%, which is far below indirect use value; the social value accounted for only 15.56%. Since the 1970s system dynamics (SD) has been widely used in the study of complex systems such as population, environment, resources, economy, and society. It is an effective tool to address nonlinear and time-lag problems. Scholars in the field have applied the SD model to study population growth, resource-carrying capacity, industrial structure, urban development, and employment problems. Liu et al. (2015) predicted China's energy consumption, total carbon dioxide emissions, and carbon dioxide emission intensity from 2013 to 2020 through SD simulation. By combining them with the energy system and renewable energy policy factors, the impacts of different economic growth rates and policy factors on energy consumption were estimated.

Momodu, Addo, Akinbami, and Mulugetta (2017) established an SD model to study the nonlinear relationship between generation adequacy and greenhouse gas emission reduction. Alamerew and Brissaud (2018) applied SD modeling to model and evaluate product recovery strategies; this was done to understand the synergistic interaction and feedback among economic, social, regulatory, environmental, managerial, and lifestyle factors in the product recovery management system. Kato, Umezu, Iwasaki, Kasanuki, and Takahashi (2013) created a development model using the SD method to study the main factors affecting the stagnation of the Japanese medical device industry and found that the stagnation could be attributed to the uncertainty associated with device development.

There are few studies studying the comprehensive benefits of marine resource development. Thus in this chapter a comprehensive benefit index system of marine resource development and utilization covering economic, social, and ecological benefits is constructed with an overall consideration of the impacts of population, social economy, resources, industrial structure, ecological services, and so on. An SD model and mean square error weight method are used to quantitatively evaluate the comprehensive benefits of marine resource utilization in 11 coastal areas in China during 2006–15. By comparing the advantages and disadvantages of marine resource development and utilization among coastal areas, we provide the basis for improving the comprehensive benefits of resource utilization.

5.3 Comprehensive benefit index system and evaluation method of marine resource utilization

5.3.1 Construction of index system

Based on research theories of marine resource development and utilization, and the principle of being objective, comprehensive, operational, and dynamic, indexes that have great impacts on the comprehensive benefits of urban marine resource development are selected and a multilevel hierarchical structure including target, category, and index layer is established. The target layer is the comprehensive benefit of marine resource development and utilization; this category level includes economic, social, and environmental benefits. The weighted average method is used to obtain weights of indexes and corresponding benefit values. The comprehensive benefit value of marine resource utilization is obtained by summing up the different benefit values. Economic benefit covers resources, manpower, technology, and capital factors; it reflects the rationalization level of marine economic structure and industrial allocation, as well as its overall ability to transform into products and services. The economic benefit indexes mainly include marine industry output value, GDP, proportion of the tertiary industry, and urbanization rate. Social benefit reflects the degree of contribution of marine resource utilization to the

Table 5.1 Evaluation indexes and weights of comprehensive benefits of marine resource utilization.

Target layer	Category level	Index layer
Comprehensive benefits of marine resources development	Economic benefit	Proportion of investment in environmental protection
		Gross output value of marine industry
		Urbanization rate
		Per capita GDP
		GDP
		Proportions of the three main industries
	Social benefit	Number of sea-related employees
		Number of scientific research personnel
		Population
		Population growth rate
		Number of tourists
		Per capita sea area
		Per capita mudflat area
	Environmental benefit	Wastewater volume
		COD emission
		Tourist capacity
		Salt pan area
		Mudflat area
		Ammonia−nitrogen emission
		Mariculture area
		Coastline length
		Wastewater treatment rate

improvement of people's livelihood in coastal areas, with the indexes mainly including the number of sea-related employees, population, and the number of scientific research personnel in the marine industry. Environmental benefit refers to the quantity and quality of marine resources in coastal areas; it indicates their ability to play a fundamental supporting role in the development of the marine economy in coastal areas. The main environmental benefit indexes are per capita mudflat area, per capita sea area, chemical oxygen demand (COD) emission, ammonia nitrogen emission, and the number of tourists. The specific index system is constructed as shown in Table 5.1.

5.3.2 Evaluation method

5.3.2.1 Standardization of indexes

In view of the different dimensions of the indexes, they should first be standardized to eliminate the impacts of dimensional inconsistencies. The indexes should be transformed to dimensionless standardized ones through appropriate transformations. There are many methods to deal with the standardization of indexes and the three common ones are the normalization, linear proportional change, and range transformation methods. In this chapter, the range transformation method is used for index standardization. After range transformation, all index values are within the range [0,1]. Usually, it can be processed according to formulas (5.1) and (5.2).

The formula for the standardization of benefit indexes is

$$Z_{ij} = \frac{\left(Y_{ij} - y_j^{\min}\right)}{\left(y_j^{\max} - y_j^{\min}\right)} \quad (i = 1, 2 \ldots n; \ m = 1, 2, \ldots m) \quad (5.1)$$

The formula for standardization of cost indexes is

$$Z_{ij} = \frac{\left(y_j^{\max} - Y_{ij}\right)}{\left(y_j^{\max} - y_j^{\min}\right)} \quad (i = 1, 2 \ldots n; \ m = 1, 2, \ldots m) \quad (5.2)$$

Z_{ij} is the standardized value of the jth index in the ith region, Y_{ij} is the original value of the jth index in the ith region; $y_j^{\max}$ and $y_j^{\min}$ are the maximum and minimum original values, respectively, of the jth index in each region.

5.3.3 Index weight

Considering the features of the marine economic system in the Bohai Sea, such as multitargets and high complexity, the demand for universality, and the stability of the comprehensive benefit evaluation index system of marine resource utilization, as well as the fact that index weighting is an important step affecting the rationality and scientificity of an index system, this chapter chooses the mean square error weighting method to determine the weights of indexes, which can effectively avoid the inherent subjectivity problem of the Delphi method, the analytic hierarchy process, and other methods. The mean square error weighting method is an important index reflecting the dispersion degree of random variables. After determining the random variables, their mean square errors are calculated, weight coefficients are obtained, and, finally, the ranking values are obtained by the multiindex decision-making method. The specific calculation formulas are shown as formulas (5.3—5.6). According to the index data of different time sections in coastal areas, these formulas are used to calculate the evaluation values of resource, economic, social, and environmental benefits of marine resource development; further, they are added up to obtain the value of the comprehensive benefit of marine resource development.

Mean value of random variables:

$$\overline{X_j} = n^{-1} \sum_{i=1}^{n} X_{ij} \tag{5.3}$$

Mean square error of random variables:

$$\sigma(X_j) = \sqrt{\sum_{i=1}^{n} \left[X_{ij} - \overline{X_j} \right]^2} \quad (i = 1, 2, \ldots n; \; j = 1, 2, \ldots m) \tag{5.4}$$

Weight of random variable:

$$w_j = \frac{\sigma(X_j)}{\sum_{j=1}^{m} \sigma(X_j)} \tag{5.5}$$

Multiindex decision–making and ranking:

$$Y_i(W) = \sum_{j=1}^{m} X_{ij} W_j \tag{5.6}$$

5.4 Construction of comprehensive benefit system dynamics model

5.4.1 Method and principle

SD, which was founded by Professor Forrest in 1956, is an interdisciplinary subject that integrates system theory and cybernetics and produces a quantitative analysis of simulated complex systems by using computer simulation technology (Anand, Vrat, & Dahiya, 2006; Ansell & Cayzer, 2018; De Stercke, Mijic, Buytaert, & Chaturvedi, 2018; Locmelis, Blumberga, Bariss, & Blumberga, 2017; Mafakheri & Nasiri, 2013). An SD model is essentially a set of differential equations with time lag, which has certain advantages in dealing with complex problems such as nonlinearity, high order, multiple feedbacks, and so on. The steps of SD modeling mainly include defining the purpose of modeling, dividing the boundary of the model, constructing the causal cycle diagram, distinguishing the nature of variables, constructing the flow chart, determining the relationship among parameters, and model checking. Model checking usually includes structural validity checking and behavioral validity checking. Only the SD model, whose validity has been checked, can be used to simulate behaviors of real systems. The SD model is built for simulation prediction and scenario analysis.

A marine ecosystem is a complex dynamic system that is influenced by many factors; further, there are nonlinear and time-lag effects among the various factors. The relationship between population, economy, society, resources, and environment should be considered systematically. Based on the feedback control theory, the SD language is used to distinguish the nature of variables, as well as between the dominant and the nondominant feedback loops. The parameters of the system state, auxiliary, and constant equations are determined with the help of Vensim platform and measurement method; based on these, the comprehensive benefit evaluation system model of marine utilization in coastal areas is constructed.

5.4.2 Structure diagram

Analyzing the relationship between the whole system and its parts is the first step to study the internal feedback structure of the system. The system structure diagram reflects the cross-linked relationship between material and information among the subsystems. By synthetically considering the mutual promotion and restriction relationship among different factors, this chapter divides the system of marine resource utilization comprehensive benefit into three subsystems: social economy, resources, and ecology. As shown in Fig. 5.1, there is a causal feedback relationship among the three subsystems. The social economy subsystem is described by regional gross national product, related marine output value, and marine industry investment. The resources subsystem is described by sea area, total volume of marine resources, and salt field area. The ecology subsystem is described by COD emission and ammonia nitrogen emission, among others.

5.4.3 Model boundary

Model boundary is a hypothetical contour that distinguishes endogenous variables from exogenous ones. Variables inside the model boundary are endogenous variables, while those outside are exogenous. All concepts

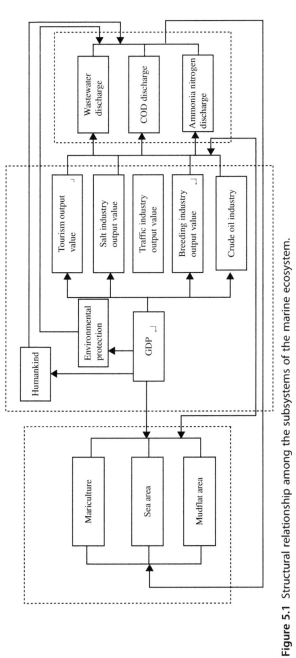

Figure 5.1 Structural relationship among the subsystems of the marine ecosystem.

and variables that are important to the problem should be considered in the model; conversely, variables without significant correlation should be excluded. Endogenous variables are often determined by the interaction among variables, whereas exogenous ones are usually determined by the external environment. In this chapter, the physical boundary of the system model is defined as the land area and offshore sea area within the administrative jurisdiction of 11 coastal areas—Liaoning, Tianjin, Hebei, Shandong, Shanghai, Guangzhou, Zhejiang, Fujian, Guangdong, Guangxi, and Hainan. The main factors affecting the comprehensive benefits of marine resource utilization include economy, society, resources, and environment. The simulation period of the SD model of comprehensive benefits of marine resource utilization is from 2006 to 2015, and the time step is 1 year.

5.4.4 Variable description

An SD model includes state variables (accumulation variables), rate variables, auxiliary variables, and constants. Among these, a state variable represents the state of a system that changes with time, and the number of state variables directly determines the complexity of the system. Rate variable is a kind of auxiliary variable that is used to describe how quickly the state variable changes over time. Usually, a rate variable is same as the number of state variable. An auxiliary variable is usually determined by other variables in the system, and its different values over time do not affect each other. A constant can be considered as a variable that does not change with time or has very little change in a certain time period. Based on previous studies and considering the availability of data, the SD model constructed in this chapter includes 5 state variables, 5 rate variables, 28 auxiliary variables, and 11 table functions. Specifically, these can be listed as follows:

State variables: population, GDP, sea area, mariculture area, and coastline length

Rate variables: variation in population, variation in GDP, variation in mariculture area, variation in sea area, and variation in coastline length

Auxiliary variables: ammonia nitrogen emission, per capita mudflat area, per capita sea area, salt pan area, marine salt industry investment, marine salt industry output value, service industry output value, tourism investment, industrial wastewater consumption, domestic wastewater volume, per capita GDP, COD emission, the proportion of marine output value, marine transportation output value, marine oil output value, other marine industries' output value, gross ocean product, marine aquaculture industry output value, capacity to handle marine tourists, number of tourists, output value of marine tourism, investment in environmental protection, mudflat area, urban population, investment in marine fisheries, and number of scientific researchers

Table functions: population change rate, GDP growth rate, proportion of the three main industries, proportion of environmental protection investment, crude oil price, urbanization rate, wastewater treatment rate, riverbank construction, riverbank natural erosion, marine transportation output value, and output value of other marine industries

5.4.5 Model construction

The statistical data needed for modeling mainly come from China City Statistical Yearbook, China Ocean Statistical Yearbook, local marine quality bulletins, provincial environmental quality bulletins, China environmental quality bulletin, Tianjin Statistical Yearbook, Dalian Statistical Yearbook, and other local yearbooks. A causal loop diagram is an intuitive tool for describing the model structure in the initial stage of model construction. According to the polarity of the loop, the causal loop can be a positive or negative feedback loop. Positive feedback enhances the deviation of variables in the loop and the trend. The system dominated by positive feedback is called a positive feedback system. Negative feedback, on the other hand, tries to stabilize the variables of the control loop and has a ground-seeking function. The system dominated by negative feedback is called a negative feedback system. Usually, the arrows between variables carry a positive or negative symbol to indicate the type of causal relationship between them. Positive symbols represent positive feedback loops and negative symbols represent negative feedback loops.

Since the causal loop diagram can only provide a simple illustration of the feedback structure within the system, it is impossible to distinguish the nature of variables; the nature of variables is further distinguished based on causality and the flow chart of the system is got by establishing the potential flow rate, so as to clearly describe the accumulative effect that affects the dynamic performance of the feedback system. In a system flow diagram, cloud-shaped symbols represent sources, representing input and output states, or all "bits." Material flow and information flow are represented by solid and dashed arrows. A flow diagram not only retains the conciseness of a causality diagram but also identifies clearly the speed and state variables. With the help of the Vensim PLE software, the flow chart of the SD model of marine ecological carrying capacity is constructed, as shown in Fig. 5.2. The main feedback loops included in Fig. 5.2 are as follows:

1. GDP → Investment in marine fisheries → Marine culture output value → Gross ocean product → GDP
2. GDP → Output value of the tertiary industry → Tourism investment → Number of tourists → Marine tourism output value → Gross ocean product → GDP
3. Population growth rate → Population size → Per capita GDP → Population growth rate
4. GDP → Per capita GDP → Number of tourists → Output value of marine tourism → Gross ocean product → GDP
5. Population size → Per capita mudflat area → Marine tourism capacity and number of tourists → Output value of marine tourism → Gross ocean product → GDP → Per capita GDP → Population growth rate → Population size

The equations of SD model include state variable equations, rate equations, auxiliary equations, constant equations, and table functions. Constants can be assigned a fixed value directly. Exogenous variables affect the endogenous variables but are not affected by them. Hence, they are often functions of time t. The model equations of different variables are mainly determined by the following methods.

1. *Empirical formula*: The variable equation can be determined by the logical relationship among the variables or by using empirical formulas.
2. *Regression analysis*: Aiming at the functional relationship among variables with high correlation, the Eviews software is used for fitting those whose average goodness of fit is above 0.9 and significant.
3. *Table function*: Usually, the nonlinear relationship among some variables is difficult to ascertain by simple linear combination. In this case, it is more convenient to show the nonlinear relationship graphically. This can be achieved by table functions in Vensim's modeling process. The model equations used in the flow chart are listed below (taking Tianjin as an example):

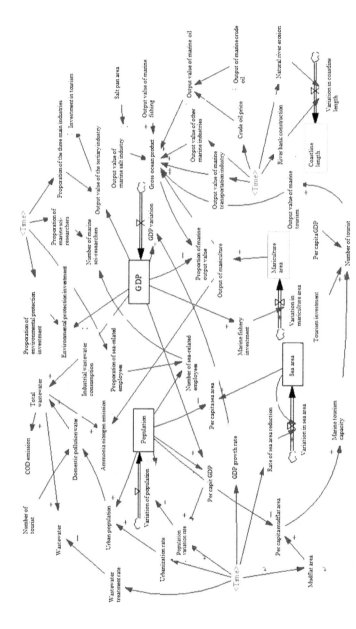

Figure 5.2 Flow chart of a comprehensive benefit evaluation system for marine resource development and utilization.

a. INITIAL TIME = 2006, representing that the base year of simulation for the model is 2006
b. FINAL TIME = 2015, indicating that the end of the simulation for the model is 2015
c. Urbanization rate = WITH LOOKUP (Time, ([(2006, 0)−(2015, 1)], (2006, 0.757), (2007, 0.763), (2008, 0.772), (2009, 0.78), (2010, 0.796), (2011, 0.805), (2012, 0.816), (2013, 0.82), (2014, 0.823), (2015, 0.826), (2016, 0)))
d. Mariculture area = INTEG (Variation of mariculture area, 5551)
e. Population size = INTEG (Variation of population, 1075)
f. Per capita mudflat area = Mudflat area/Population size
g. Total amount of wastewater = Industrial wastewater consumption + Domestic pollution water
h. Gross ocean product = Mariculture output value + Marine oil output value + Marine transportation output value + Output value of other marine industries + Marine fishery output value + Marine tourism output value + Marine salt industry output value
i. Output value of the tertiary industry = GDP × The proportion of the tertiary industry
j. Environmental protection investment = GDP × The proportion of environmental protection investment
k. Wastewater discharge = Total wastewater × Wastewater treatment rate
l. The proportions of three main industries = WITH LOOKUP (Time, ([(2006, 0)−(2015, 1)], (2006, 0.402), (2007, 0.405), (2008, 0.379), (2009, 0.453), (2010, 0.46), (2011, 0.462), (2012, 0.47), (2013, 0.481), (2014, 0.496), (2015, 0.522)))
m. GDP = INTEG (GDP variation, 4462.74)
n. Urban population = Population size × Urbanization rate
o. Marine oil output value = Crude oil price × Marine oil output
p. Sea area = INTEG (Variation of sea area, 171,800)
q. Number of sea-related employees = Proportion of sea-related employees × Population size
r. Number of marine scientific researchers = Proportion of marine scientific researchers × Number of sea-related employees
s. Wastewater treatment rate = WITH LOOKUP (Time, ([(2000, 0)−(2015, 1)], (2006, 0.97), (2007, 0.971), (2008, 0.975), (2009, 0.99), (2010, 0.974), (2011, 0.982), (2012, 0.992), (2013, 0.982), (2014, 0.987), (2015, 0.989)))

5.5 Comprehensive benefit evaluation results and analysis

5.5.1 Model test

5.5.1.1 Validity test

The validity tests of the model include the structural validity test and behavioral validity test. Structural validity test mainly tests the rationality of dimension, variable properties, and equation setting. Behavioral validity test is to verify whether the model can replicate the established behavior pattern of the system (Du, Li, Zhao, Ma, & Jiang, 2018; Wu et al., 2017). The model constructed in this chapter has passed the dimension consistency test and equation structure test available on the Vensim software platform; this shows that the model has certain validity in terms of structure. The historical and simulation data of the main variables in three different subsystems (taking Tianjin as an example), that is, per capita GDP, fishing output, and salt pan area from 2006 to 2015, are selected to test the validity of the parameters. From Table 5.2, it can be seen that the relative errors between the historical value and the simulation value of the three important indexes are within 5%; this means that the data coincidence degree meets the requirement of parameter validity and the model can be used to simulate the behavior pattern of the real system owing to its high replication ability.

5.5.1.2 Stability test

The behavior criterion of SD models is "structure determining function." It is emphasized that the behavior mode of the system is mainly determined by the internal dominant feedback loop, and the change in parameters in the nondominant loop usually does not cause significant changes in the system's behavior mode (Bautista, Espinoza, Narvaez, Camargo, & Morel, 2019; Iandolo, Barile, Armenia, & Carrubbo, 2018; Zetterberg, 2014). In other words, an effective system model should be insensitive to most parameter changes. In this paper, GDP is used as a test variable, and the time step is set as 12 months, 6 months, and 3 months to observe whether there is any change in the trends observed for the variables. As shown in Fig. 5.3, the trends of the tested variables are almost the same under the three time steps; this means that the model is

Table 5.2 Validity test of parameters.

Year	Per capita GDP (CNY)			Fishing output (t)			Salt pan area (ha)		
	Historical value	Simulation value	Relative error (%)	Historical value	Simulation value	Relative error (%)	Historical value	Simulation value	Relative error (%)
2006	41,163	42,293	2.745	32,827	33,892	3.244	33,649	32,503	− 3.406
2007	46,122	46,122	0.000	30,185	31,178	3.290	33,397	31,960	− 4.303
2008	55,473	55,929	0.822	24,494	23,877	− 2.519	30,788	31,771	3.193
2009	62,574	60,189	− 3.811	16,459	16,362	− 0.589	30,866	31,223	1.157
2010	72,994	73,611	0.845	15,754	15,098	− 4.164	29,050	30,126	3.704
2011	85,213	86,245	1.211	13,305	13,808	3.781	27,461	27,901	1.602
2012	93,173	94,376	1.291	14,285	14,901	4.312	26,470	27,564	4.133
2013	100,105	100,449	0.344	53,437	53,774	0.631	26,499	26,844	1.302
2014	105,231	107,110	1.786	45,548	46,932	3.039	26,242	24,980	− 4.809
2015	107,960	108,256	0.274	47,094	46,968	− 0.268	26,221	27,141	3.509

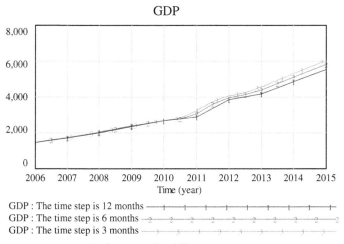

Figure 5.3 Variation trends of GDP under different time steps.

insensitive to a change in time steps. Because the structure and function of the model are stable, it can be used to simulate the behavior of real systems.

5.5.2 Simulation results

The development and utilization of marine resources are influenced by many factors, such as economy, society, resources, ecology, and the environment. The comprehensive benefits of marine resource utilization vary across regions and time periods. Fig. 5.4 shows the spatial and temporal evolution characteristics of comprehensive benefits of marine resource utilization in 11 coastal areas of China.

It can be seen from Fig. 5.4 that the comprehensive benefit level of marine resource development and utilization in coastal areas of China has, overall, maintained an upward trend, but the growth rates and growth trends vary across regions. In terms of growth rate, the comprehensive benefit indexes of six areas—Tianjin, Shandong, Shanghai, Jiangsu, Liaoning, and Guangxi—have risen rapidly in the past 10 years, with an average annual growth rate of more than 2%; Shandong province had the highest growth rate of 2.92%. This shows that these regions have fully utilized the advantages of economy, policy, and location to register a remarkable improvement in their utilization rates of marine resources. The comprehensive benefits in Hebei, Fujian, and Guangdong have shown a steady trend, with an average annual growth rate of 1.7%.

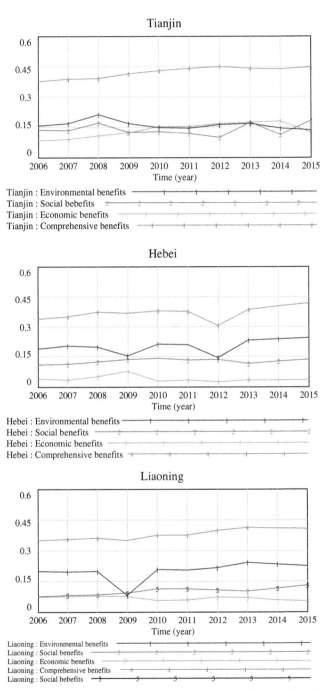

Figure 5.4 Spatial and temporal evolution characteristics of comprehensive benefits of marine resource utilization.

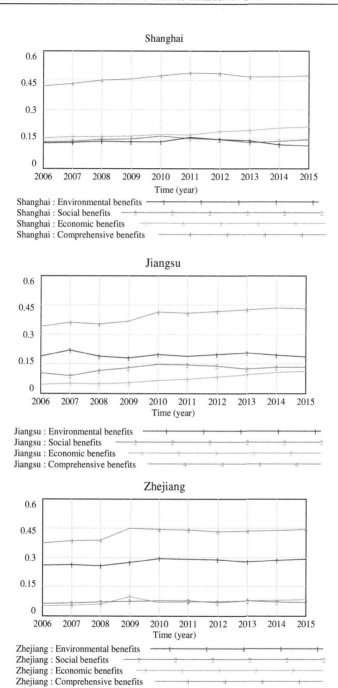

Figure 5.4 (Continued)

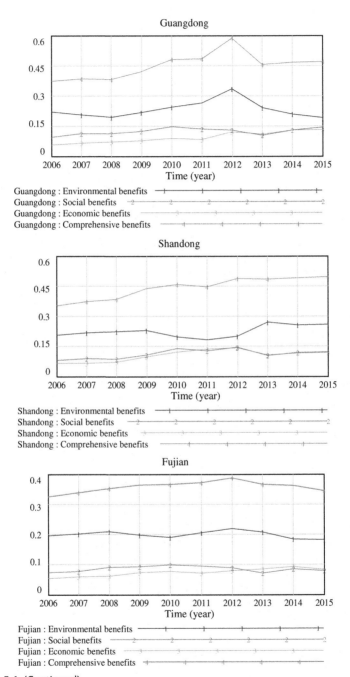

Figure 5.4 (Continued)

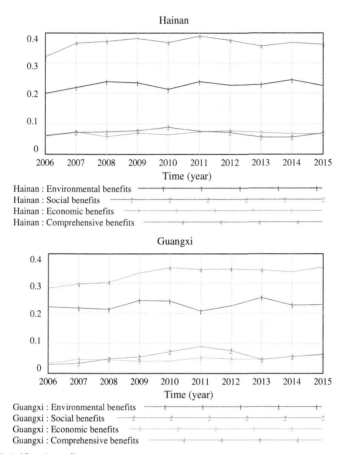

Figure 5.4 (Continued)

This shows that the development and utilization of marine resources are proceeding in an orderly manner; however, the development efficiency can be further improved. The comprehensive benefits of marine utilization in Hainan and Zhejiang provinces had the lowest increase, with an average annual growth rate of about 1%. This shows that their utilization efficiency of marine resources, as well as the degree of coordination among the regional economy, resources, and the environment, was low. These areas urgently need relevant development policies to standardize the development behavior of coastal marine industry, reduce the negative impact of environmental pollution, and promote the comprehensive benefits of regional marine resources. Judging from the comprehensive benefit value of marine resource development and utilization, the descending ranks of the provinces were Shanghai, Guangdong, Zhejiang, Shandong, Tianjin, Jiangsu, Liaoning, Hebei, Fujian, Hainan, and Guangxi; in 2015 the order was Shanghai, Shandong, Guangdong, Jiangsu, Tianjin, Hebei, Liaoning, Zhejiang, Fujian, Hainan, and Guangxi. From the change of ranking in the research period, it can be seen that the comprehensive benefits of marine resource utilization in Shanghai, Guangdong, Jiangsu, Tianjin, and Shandong have been above the average value for all coastal areas, and the overall ranking is basically stable. The rank of Zhejiang province dropped dramatically, while those of Hebei, Fujian, Hainan, and Guangxi remained stable.

The level of comprehensive benefits of marine resource utilization depends on the level of economic, social, resource, and environmental benefits. The economic benefits of marine resources in all regions showed a rising trend. In 2006 the range of economic benefits of marine resource utilization in different regions was [0.042−0.079], while in 2015 the range was [0.06−0.22]. The highest growth occurred in Tianjin, with an average annual growth rate of 10.03%. Shanghai, Tianjin, Jiangsu, Shandong, and Guangdong are provinces where the growth rate was higher than the average annual growth rate. This shows that the marine industrial structure of these areas was more suitable; further, the advantages of marine industries were fully exploited and there was a huge improvement in economic benefits.

From the perspective of social benefits of marine resource utilization, Shandong had the highest average annual growth rate of 2.68%. Growth rates in Shandong, Guangdong, Liaoning, Tianjin, and Shanghai were higher than the average annual growth rate. In 2006 the range of social benefits of marine resource utilization in different regions was [0.04−0.121], while in 2015 the range was [0.052−0.151]. Higher social benefits mean that the level of employment, education, and scientific research in the region improved with the development of marine economic industry.

From the perspective of environmental benefits of marine resource utilization, Shandong province had the highest average annual growth rate of 2.92%. The provinces where the growth rates were higher than the average annual growth rate are Shandong, Hebei, Liaoning, Guangxi, and Hainan. This means that these provinces paid more attention to the protection of marine ecological environment while enhancing the development of marine resources. In 2006 the range of environmental benefits of marine resource utilization in different regions was [0.144−0.248], while that in 2015 was [0.143−0.388].

5.6 Conclusions and suggestions

5.6.1 Conclusions

By comparing the advantages and disadvantages of marine resource development and utilization among coastal provinces and cities horizontally, this research provides the basis for improving the comprehensive benefits of resource utilization. Considering the influences of population, social economy, resources, industrial structure, and ecological service, this chapter establishes a comprehensive benefit evaluation index system covering economic, social, and environmental benefits. The comprehensive benefits of marine resource utilization in 11 coastal areas of China from 2006 to 2015 are quantitatively evaluated by building a comprehensive benefit evaluation model. The following conclusions are drawn.

(1) During the modeling phase, the comprehensive benefit system model of marine resource development and utilization constructed in this chapter passed the consistency tests of model structure and dimension available on the Vensim PLE software. By selecting representative variables for the historical test, it is shown that the relative error between the simulation and historical values of parameters is less than 5%, which means that the behavior of the system can be effectively replicated. Besides, there are no substantial changes in the variation trends of the selected parameters under different time steps, indicating that the model has certain stability. Therefore the SD model of comprehensive benefit evaluation of marine resource utilization can effectively simulate the behavior modes of real systems.

(2) Over the past decade, the comprehensive benefits of marine resource utilization in coastal areas in China showed a fluctuating growth trend. The comprehensive benefit levels in different regions varied across time. The levels of economic, social, and environmental benefits in each region also showed a growth trend. In 2005 the comprehensive benefit value of 11 coastal areas was in the range [0.29−0.42], and in 2015 the range was [0.36−0.49].

(3) The comprehensive benefit level of marine resource utilization depends on the economic, social, and environment benefits. Provinces with high comprehensive benefit values may not have had balanced development in all three aspects. The high level of benefits

caused by the high-intensity development may have weakened the environmental protection consciousness. Balancing the contradictory demands of the economy, society, and environment to maximize the comprehensive benefit value is a problem that government departments at all levels need to consider carefully; further, it is also an urgent problem requiring resolution if the sustainable development of marine undertakings is to be maintained. According to the results of benefit evaluation of different coastal areas, appropriate strategic measures should be taken to coordinate and balance the relationship among economic, social, and environmental benefits to achieve the promotion of all three.

(4) There are significant differences in the comprehensive benefit levels of marine resource utilization among the 11 coastal areas of China. Judging from the situation in 2015 the descending order of ranking was Shandong, Shanghai, Guangdong, Tianjin, Zhejiang, Jiangsu, Hebei, Liaoning, Hainan, Guangxi, and Fujian. It is found that areas with high comprehensive benefits do not have characteristics of spatial agglomeration and their marine comprehensive benefits were higher than those of their surrounding areas. There was negative spatial correlation between adjacent areas. Thus the advantages of resources, basic economic conditions, and scientific and technological strength should be fully exploited; the construction of marine comprehensive development pilot zones should be quickened; and new industries should be fostered with an emphasis on enhancing the desirable effects of marine industries on surrounding areas.

5.6.2 Suggestions

1. Strengthen the construction of marine nature reserves

Frequent occurrence of marine disasters is a common problem faced by coastal areas in China, especially in Fujian and other areas prone to marine disasters. The social and economic losses caused by marine disasters have become one of the important factors restricting the efficiency of marine resource utilization in coastal areas of China, especially in areas with long coastlines and dense population. Ecological processes and genetic resources are two important aspects to consider when ensuring the sustainable development of marine ecosystem. On the one hand, marine nature reserves can maintain ecosystem productivity and important ecological processes by controlling disturbance and physical destruction activities. On the other hand, the number of organisms can be increased by strengthening the conservation and reproduction of marine living resources to establish natural habitats for endangered animals that are protected. Meanwhile, the relevant legislation on nature reserves should be improved and all kinds of illegal acts that undermine the construction of marine natural protection should face a strict crackdown.

2. Pay equal attention to proper development and environmental protection

    The rapid development of the social economy and the exploitation and utilization of marine resources promote and restrict each other. Appropriately coordinating the relationship among the marine economy, society, resources, and the environment is conducive to the healthy and sustainable development of the marine economy. A one-sided pursuit of high economic growth and the neglect of marine ecological environment construction should be avoided. Meanwhile, an excessive pursuit of marine ecological environment quality should also be eschewed because it may restrain the development of the marine industrial economy. The exploitation and utilization of resources and the protection of ecological environment need to develop harmoniously to promote a virtuous circle of marine economic development.

3. Create a low-carbon marine industry economic demonstration zone

    With the rapid development of the low-carbon economy, it is imperative to upgrade and adjust the marine low-carbon industrial structure. To give full play to the demonstration effect of low-carbon marine economic demonstration zone, on the one hand, the development and investment of marine new energy industry should be accelerated and the mode of economic growth should be changed. On the other hand, the rise and development of modern marine services, such as coastal tourism, modern marine logistics, and marine financial industry, should be quickened to occupy the commanding heights of marine low-carbon industries. The demonstration effect of the low-carbon marine industry economic demonstration zone can guide the development and investment of the marine low-carbon industry in energy-saving and environmental protection projects; further, the application of low-carbon technology to the marine industry can lead to a benign interaction among marine resources, ecology, and the environment.

4. Strengthen investment in marine science and technology to optimize and upgrade industrial structure

Strengthening investment in marine science and technology to improve the technological level of marine resources development is the only way to realize the upgrading of marine industrial structure and the transformation of the marine economic growth mode. On the one hand, the communication and cooperation between industry, universities, and research institutes should be strengthened in the marine field to build first-class marine science and technology research institutions and ocean universities and innovate the training of talents engaged in marine science and technology. On the other hand, domestic and international cooperation and exchanges in marine science and technology should be enhanced to introduce new technologies, achievements, and technologies, as well as improve the scientific and technological capabilities for marine development. According to the geographic and economic characteristics of different marine regions and the needs of local social and economic development, the layout of the marine industry should be adjusted to promote the optimization and upgradation of the marine structure.

5. Control the emission of pollution sources and restore the function of the marine ecological service

Controlling the discharge from pollution sources is an important task in improving the function of marine ecological services. With the acceleration of the urbanization process, the direct discharge of water from domestic and industrial pollution destroys the cyclic development of the marine ecology. Effective control of discharge from pollution sources requires the construction of treatment plants for domestic and functional water use in cities and towns to improve the standard rate of sewage discharge, as well as the provision of corresponding antifouling equipment and materials to prevent the marine pollution caused by ports, ships, and offshore oil platforms.

6. Control the population size and relieve the pressure on ecological resources

Human activity is the ultimate source of environmental pollution and waste of resources. Strict control of population growth is the only way to alleviate the pressure on resources and the environment and improve the carrying capacity of the ecological environment. The relative insufficiency of resources per capita in coastal areas highlights the necessity and urgency of controlling population growth. Controlling the rapid growth of population can, on the one hand, reduce the demands for resources arising out of human activities; on the other hand, the increase of per capita resources can enhance, to a certain extent, the social benefits of marine resource utilization.

7. Adapt to time and local conditions

Because of different types of resource advantages, economic foundations, and policy guarantees, coastal areas should adopt development strategies that consider local conditions to develop and utilize their marine resources. Moreover, a good sense of marine resources protection should be established; further, the management level of marine resource development and utilization should be improved and the order of marine resource development should be standardized to enhance the comprehensive competitiveness of China's marine resources. Areas with the potential of high comprehensive benefits should give full play to their resource advantages, basic economic conditions, and scientific and technological strength. They should make full use of the demonstration effect of marine comprehensive development pilot zones, focus on fostering new marine industries, and enhance the driving effect of demonstration zones on surrounding areas.

References

Alamerew, Y. A., & Brissaud, D. (2018). Modelling and assessment of product recovery strategies through systems dynamics. *Procedia CIRP, 169*, 822−826.

Alfonso, A. M., Robert, W. B., Victor, C. I., Jose, D. C., Silvia, E. I. O., Barbara, W. M., ... Fredrik, W. (2001). Sustainability of coastal resource use in San Quintin, Mexico. *AMBIO: A Journal of the Human Environment, 30*(3), 142−149.

Anand, S., Vrat, P., & Dahiya, R. P. (2006). Application of a system dynamics approach for assessment and mitigation of CO_2 emissions from the cement industry. *Journal of Environmental Management, 79*, 383−398.

Ansell, T., & Cayzer, S. (2018). Limits to growth redux: A system dynamics model for assessing energy and climate change constraints to global growth. *Energy Policy, 120*, 514−525.

Bautista, S., Espinoza, A., Narvaez, P., Camargo, M., & Morel, L. (2019). A system dynamics approach for sustainability assessment of biodiesel production in Colombia: Baseline simulation. *Journal of Cleaner Production, 213*, 1−20.

Cavanagh, R. D., Broszeit, S., Pilling, G. M., Rant, S. M., Murphy, E. J., & Austen, M. C. (2016). Valuing biodiversity and ecosystem services: A useful way to manage and conserve marine resources? *Proceedings of the Royal Society B: Biological Sciences, 283* (1844), 1−8.

Cognetti, G., & Maltagliati, F. (2010). Ecosystem service provision: an operational way for marine biodiversity conservation and management. *Marine Pollution Bulletin, 60*(11), 1916.

Costanza, R. (1999). The ecological, economic, and social importance of the oceans. *Ecological Economics, 31*(2), 199−213.

Daniel, M., Andrew, C., Ian, J., & Jin, W. (2014). A model assessing cost of operating marine systems using data obtained from Monte Carlo analysis. *Proceedings of the Institution of Mechanical Engineers, Part M: Journal of Engineering for the Maritime Environment, 288*(4), 398−412.

De Stercke, S., Mijic, A., Buytaert, W., & Chaturvedi, V. (2018). Modelling the dynamic interactions between London's water and energy systems from an end-use perspective. *Applied Energy, 230*, 615−626.

Duan, X. F., & Xu, X. G. (2009). Regional differences evaluation on the comprehensive benefit of marine resources development and utilization. *Acta Scientiarum Naturalium Universitatis Pekinensis, 45*(6), 1055−1060.

Du, L. L., Li, X. Z., Zhao, H. J., Ma, W. C., & Jiang, P. (2018). System dynamic modeling of urban carbon emissions based on the regional National Economy and Social Development Plan: A case study of Shanghai city. *Journal of Cleaner Production, 172,* 1501−1503.

Fathab, B. D. (2015). Quantifying economic and ecological sustainability. *Ocean and Coastal Management, 108,* 13−19.

Francisco, C. P., Rosa, G. F., Francisco, A. S., & Carlos, F. G. (2014). Using stakeholders' perspective of ecosystem services and biodiversity features to plan a marine protected area. *Environmental Science and Policy, 40,* 116−131.

Fu, X. M., Wang, N., & Jiang, S. S. (2018). Value evaluation of marine bioresources in Shandong offshore area in China. *Ocean & Coastal Management, 163,* 296−303.

Gao, S., Sun, H. H., Zhao, L., Wang, R. J., Liu, B. Q., & Xu, M. (2017). Comprehensive benefit evaluation of various types of marine development: a case study of Jiangsu. *Fresenius Environmental Bulletin, 26*(10), 6175−6183.

Giselle, S. T., Alan, W., Mary Ann, T., John, D., Esperanza, E., & Ciemon, C. (2007). Economic valuation of coastal and marine resources: Bohol Marine Triangle, Philippines. *Coastal Management, 35*(2-3), 319−338.

Iandolo, F., Barile, S., Armenia, S., & Carrubbo, L. (2018). A system dynamics perspective on a viable systems approach definition for sustainable value. *Sustainability Science, 13* (5), 1245−1263.

Kato, T., Umezu, M., Iwasaki, K., Kasanuki, H., & Takahashi, Y. (2013). Preliminary study on the development of a system dynamics model: the case of EVAHEART. *Journal of Artificial Organs, 16*(2), 242−247.

Liu, X., Mao, G. Z., Ren, J., Li, R. Y. M., Guo, J., & Zhang, L. (2015). How might China achieve its 2020 emissions target? A scenario analysis of energy consumption and CO2 emissions using the system dynamics model. *Journal of Cleaner Production, 103,* 401−410.

Locmelis, K., Blumberga, A., Bariss, U., & Blumberga, D. (2017). Energy policy for energy intensive manufacturing companies and its impact on energy efficiency improvements. *System dynamics approach. Energy Procedia, 128,* 10−16.

Mafakheri, F., & Nasiri, F. (2013). Revenue sharing coordination in reverse logistics. *Journal of Cleaner Production, 59,* 185−196.

Momodu, A. S., Addo, A., Akinbami, J. F. K., & Mulugetta, Y. (2017). Low-carbon development strategy for the West African electricity system: preliminary assessment using system dynamics approach. *Energy Sustainability & Society., 7*(1), 11.

Paul, R., & Teresa, F. (2003). Management of environmental impacts of marine aquaculture in Europe. *Aquaculture, 226*(1), 139−163.

Pontecorvo, G., & Wilkinxon, M. (1980). Contribution of the Ocean Sector to the U.S. Economy. *Science, 208,* 1000−1006.

Porobic, J., Fulton, E. A., Frusher, S., Parada, C., Haward, M., Ernst, B., & Stram, D. (2017). Implementing ecosystem-based fisheries management: Lessons from Chile's experience. *Marine Policy, 97,* 82−90.

Wei, T., Li, S. J., & Zhao, P. (2012). Appraisal on the comprehensive benefits of marine resources development and utilization in China. *Environmental Protection Science, 38*(4), 81−84.

Wu, Y. B., Li, M. N., Liu, L., Zhang, Y. F., Liu, L., & Wang, L. (2017). Spatial-temporal allocation of regional land consolidation project based on landscape pattern and system dynamics. *Cluster Computing, 20*(4), 3147−3160.

Zetterberg, L. (2014). Benchmarking in the European Union emissions trading system: Abatement incentives. *Energy Economics, 43,* 218−224.

CHAPTER SIX

Decoupling marine resources and economic development in China

6.1 Introduction

Since the 1990s, the world's marine economy has been developing rapidly and in coastal countries, its economic impact has been increasing. With the growing population pressure and the continuous consumption of land resources, oceans with rich resources have gradually become the focus of strategic development in various countries. At present, the development of marine resources is still in its infancy. Due to the limitation of development technology and the imperfection of relevant laws and regulations, the degree of marine resource exploitation is still low. There are also various problems in the process of development.

As an increasing number of coastal countries attach importance to the development of their marine economy, the exploitation and utilization of marine resources have gradually become the focus of the sustainable development in many national economies. Coastal countries continuously make decisions and plans for the strategic advancement of their marine resources to achieve their rational utilization and in-depth development. As an important basis for the development of the blue economy, the development of marine resources has been accelerating worldwide, and the scale of development has been expanding. Concurrently, the development and utilization of marine resources are playing an increasingly important role in the sustainable development of social economies in coastal countries.

As a country with extensive marine development and utilization, China's marine economy is evolving rapidly and has become an important component of its national economy. Against the backdrop of breakthroughs in marine technology and increasingly fierce competition in the world, China's reliance on the economic development of the ocean has risen to a new height. After a period of rapid marine economy development, China is now in a transitional period of high quality and steady marine development. During this period, the driving factors of marine economic growth are

Sustainable Marine Resource Utilization in China.
DOI: https://doi.org/10.1016/B978-0-12-819911-4.00006-0

gradually changing. Meanwhile, various problems have emerged in the transformation process of old and new industries. Overfishing in marine fisheries, aggravation of marine pollution, and constant breakdown of the marine ecological balance have attracted increasing attention.

To meet the new needs of marine economic development and the new structure of sustainable development of the marine economy, marine economic development needs to eliminate gradually excessive dependence on marine resources. On the one hand, the driving factors of marine economic development should be reformed, and the marine industrial structure should be optimized and upgraded. On the other hand, the consumption of marine resources should be lowered to achieve a gradual decoupling of marine economic development and marine resources. In this chapter, the existing achievements of scholars are reviewed and the overall trend of China's marine economic growth and total resource consumption is analyzed based on the concepts of the marine ecological footprint, the decoupling of marine economic growth and resource consumption, and a quantitative estimation of China's marine ecological footprint. In addition, a model of the factors affecting the decoupling of marine resources is established, which reveals the influencing factors of the decoupling state of marine resource utilization in-depth, and finally, corresponding countermeasures and suggestions are put forward based on the empirical results.

6.2 Literature review

There are complex interactions and relationships between the marine economy and marine resources. Development of the marine economy is accompanied by the consumption of its resources. As resource consumption becomes part of the bottleneck of development technology and reserves, the development of the marine economy is restricted.

In an earlier period, extending the marine economy depended on factor inputs, where a high output of marine products could only be realized through an increase in marine resource inputs under the premise of an unchanged technological level and marine personnel. When the marine economy developed to a certain extent, through technological progress in the marine industry and continuous adjustment and optimization to its structure, the economic shares of primary and secondary industries began to decline, and the mode of marine economic growth began to intensify. In this stage, due to the improvement in production efficiency, the same output in the past needed a smaller amount of marine resource input, thereby reducing the pressure of marine economic development on marine resources. Generally, although technological progress in this stage significantly improved the utilization efficiency of marine resources and further reduced their consumption, the improved utilization efficiency also led to a lower cost for marine resources and, consequently, increased consumption. Thus the growth in the marine economy leads to an increase in marine resource consumption and marine resources and the savings through technological progress are thereby offset to some extent (Figge, Young, & Barkemeyer, 2014).

There have been many studies on the relationship between marine economic development and resource consumption, mainly from three perspectives.

1. The role of marine resources in the marine economy. Side and Jowitt (2002) predicted the future direction of marine resource utilization. They believed that strengthening marine resource management would be an important driving force in marine technological progress. Meanwhile, effective systems, such as reserving material habitats, would need to be established to ensure the sustainable development of marine resources. Guillaumont (2010) measured the adaptability of marine resources in small island economies by building a vulnerability assessment framework. Li, Yang, and Su (2016) used set pair analysis to analyze the evolution trend of China's marine economic vulnerability, pointing out that the rational use of marine resources is an effective way to reduce vulnerability.

2. The role of the marine economy in marine resources. Managi, Opaluch, Jin, and Grigalunas (2005) explored the impacts of technological progress on oil and gas development in the Gulf of Mexico over the past 50 years using micro-data; the results showed that technological progress could increase the reserves of marine resources, reduce their consumption, and offset the phenomenon of resource depletion. Samonte-Tan, White, and Tercero et al. (2007) determined the values of various resources by calculating the net benefits generated by them in the marine triangle in the Baohe Island in the Philippines, and explored the correct coastal service and management modes. Gogoberidze (2012) estimated the potential of the marine economy and applied it to coastal zone planning.

3. The interaction between the marine economy and the marine environment. Halpern, Longo, and Hardy (2012) comprehensively and

quantitatively measured the health status of the man—sea coupling system in coastal countries by constructing a man—sea relationship index. Barange, Cheung, Merino, and Perry (2010) constructed a coordination model between the development of the marine economy and the development of marine resources to explore their sustainable development path. Sun, Zhang, Zou, and Zeyu (2015) used the analytic hierarchy process and projection pursuit model and information entropy model to evaluate the social and resource status of coastal areas, and explored the coastal direction of the man—sea relation.

The above studies either preferred to analyze the composition of the total marine economy and explore the complex relationship between the development of the marine economy and the exploitation of marine resources by means of traditional economic quantitative methods or conducted qualitative research on the relationship and mechanism of the marine economy and resources from the perspective of economic development and ecological evolution. Therefore these studies lack convincing quantitative results on the decoupling relationship between marine resources and the marine economy and its influencing factors. Moreover, often, previous research results have been about temporal or spatial variations in the same region, and thus lack a comprehensive investigation of spatial and temporal scales. To address these gaps, this study applies a revised marine ecological footprint to reflect the consumption of resources in marine economic activities. It then scientifically evaluates the coordinated relationship between marine economic growth and resource consumption in China's coastal areas using a decoupling evaluation model of coordinated development to provide a scientific basis and a decision-making reference for realizing the sustainable development of marine resource utilization and marine economic growth in coastal areas.

6.3 Evaluation of the marine ecological footprint

6.3.1 Determining the marine ecological footprint and equilibrium factors

The ecological footprint was first proposed by Rees (1992) to calculate the demands of human production activities for natural productive lands. By converting various resources and the energy consumed by human activities into certain proportions, the corresponding bio-productive land area (Monfreda, Wackernagel, & Deumling, 2004) needed to supply these resources and energy was obtained. As an effective method to calculate environmental consumption, the ecological footprint can quantitatively measure the extent of resource consumption. The marine ecological footprint refers to the surface area of the marine ecosystem used by humans to meet their needs for seafood and other services (Wernberg, Thomsen, & Connell, 2013) based on unchanged technological conditions. Many reasons have led to the high complexity of the research on the marine ecological footprint such as the mobility of the marine water body, the complexity of the marine environment, the public ownership and renewability of marine resources, and the common three-dimensional exploitation of marine resources.

According to the characteristics of oceans and referring to the relevant literature (Jones, Doblin, Matear, & King, 2015), this chapter presents a revised marine ecological footprint model. Based on the basic functions of different sea areas, productive marine land can be divided into three types:

1. Sea fishery area: mainly used for the cultivation and fishing of marine fish, shrimp, shellfish, algae, and so on, measured by the number of fish caught and the amount of cultivation of marine products (Fock, Kloppmann, & Probst, 2014);

2. Sea energy area: mainly used for the production and development of the marine chemical industry and for the exploitation of marine mineral oil and gas resources, including the coastal chemical industry, mining of mineral resources, exploitation of oil and gas resources, and wind energy.

3. Mudflat area: mainly used for the sea salt industry and the mudflat cultivation industry, and measured by industrial output.

The determination of the equilibrium factor: as different types of land have different ecological productivity, to compare and calculate the different productivity of these different types of land, we need to convert these land areas into areas with the same biological productivity according to certain standards. The conversion coefficient used is called the equilibrium factor. In the past, in the widely used ecological footprint model, the equilibrium factor has been mainly based on the weight of land-based productivity, while ignoring the weight of long-term technological potential and other economic and social aspects. Moreover, the general fixed values have been used often which cannot show the spatial and temporal differences of different land types. In view of this, a revised global

entropy method to determine the equilibrium factor of the marine ecological footprint is used here. The global entropy method can include different information from each index in the process of determining the
index weight; thus it can comprehensively and objectively reflect the relative biological productivity of different types of marine land. The calculation is divided into four steps as follows.

1. Carry out standardization and coordinate translation of different indices to ensure that all standardized values are positive.
2. Calculate the ratio R_{ij} of X_{ij}', where,

$$R_{ij} = X_{ij}' / \sum_{i=1}^{m} X_{ij}'$$

3. Calculate the entropy value e_j of the jth index, where
$e_j = -\frac{1}{\ln m} \sum_{i=1}^{m} R_{ij} \ln R_{ij}, \ 0 \le e_j \le 1;$
4. Calculate the weight w_j of index x_j, and the expression is $w_j = \dfrac{1 - e_j}{\sum\limits_{j=1}^{n}(1 - e_j)}$.

6.3.2 Spatial–temporal evolution analysis of the marine ecological footprint

As shown in Table 6.1, from 2008 to 2015, the marine ecological footprint
of coastal areas mainly consists of the marine footprint and the energy ecological footprint, while the proportion of the mudflat ecological footprint is relatively small. The marine energy ecological footprint accounts for the largest
proportion of the marine ecological footprint at more than 70%, mainly
because of favorable local conditions such as flat terrain, climate, population,
and transportation. Table 6.1 shows that the fishery ecological footprint in
coastal areas fluctuated between 2008 and 2015, but was generally stable. The
coastal mudflat ecosystem maintained a similar trend, generally stable at about
9×10^5 hectares. The energy ecological footprint showed an upward trend
year by year, which indicates that human activities in China's coastal areas
caused increasing pressure on the ocean to absorb CO_2. The marine ecological footprint in coastal areas showed an upward trend year by year, and the
growth of the energy ecological footprint was the main reason.

Fig. 6.1 shows that similar to the marine ecological footprint, except in
2013, the per capita marine ecological footprint in coastal areas showed an overall upward trend, as did the per capita energy ecological footprint. However,
the per capita fishery footprint gradually declined between 2008 and 2010, and
showed a slow upward trend after reaching the lowest level in 2010.

Table 6.1 Marine ecological footprints in coastal areas (hm^2).

Marine ecological footprint types	2008	2009	2010	2011	2012	2013	2014	2015
Fishery footprint	4.68×10^8	4.69×10^8	4.21×10^8	4.34×10^8	4.44×10^8	4.43×10^8	4.50×10^8	4.62×10^8
Mudflat footprint	9.18×10^5	9.45×10^5	9.67×10^5	9.95×10^5	9.56×10^5	9.40×10^5	9.39×10^5	8.92×10^5
Energy footprint	1.27×10^9	1.36×10^9	1.51×0^9	1.66×10^9	1.70×10^9	1.69×10^9	1.72×10^9	1.77×10^9
Marine footprint	1.75×10^9	1.83×10^9	1.93×10^9	2.01×10^9	2.15×10^9	2.14×10^9	2.18×10^9	2.23×10^9

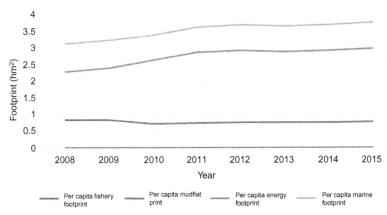

Figure 6.1 Variations in per capita marine ecological footprints in coastal areas.

As for each coastal area (Table 6.2), based on 2015, Shandong Province consumed the largest marine ecological footprint, reaching 4.77×10^8 hectares in 2015. This was related to its abundant marine resources, its long history of development, the intensive development and utilization of sea areas and coastal zones, and its developed secondary industry. Following Shandong, the marine ecological footprints of the coastal areas from large to small were in order, Jiangsu, Liaoning, Guangdong, Zhejiang, Hebei, Fujian, Shanghai, Guangxi, Hainan, and Tianjin. The order of the per capita marine ecological footprints from large to small was Hainan, Liaoning, Shandong, Zhejiang, Tianjin, Fujian, Shanghai, Jiangsu, Hebei, Guangdong, and Guangxi. Although the marine ecological footprints of Hainan and Tianjin were smaller than those of other coastal areas, their per capita marine ecological footprints were larger, which means that the individual consumption of marine resources was high there even though the absolute consumption of marine resources was small.

The marine ecological footprints of Tianjin, Jiangsu, Zhejiang, Fujian, Shandong, and Hainan showed a generally upward trend year by year. The pressure on the oceans caused by these areas has increased continuously. The marine ecological footprint of Hebei Province increased year by year up to 2012, and then began to show a downward trend after reaching a maximum of 2.50×10^8 hectares in 2012. This trend resulted mainly from a decline in its energy ecological footprint. Liaoning Province showed a similar trend with its highest marine ecological footprint occurring in 2012 before a downward trend and eventual

Table 6.2 Marine ecological footprints (hm^2) of coastal areas.

Total footprint	2008	2009	2010	2011	2012	2013	2014	2015
Tianjin	4.31×10^7	4.63×10^7	5.88×0^7	6.47×10^7	6.47×10^7	6.85×10^7	6.64×10^7	6.75×10^7
Hebei	1.92×10^8	2.02×10^8	2.19×10^8	2.47×10^8	2.50×10^8	2.49×10^8	2.39×10^8	2.41×10^8
Liaoning	2.37×10^8	2.43×10^8	2.47×10^8	2.63×10^8	2.75×10^8	2.63×10^8	2.63×10^8	2.59×10^8
Shanghai	8.98×10^7	8.95×10^7	9.24×10^7	9.50×10^7	9.56×10^7	1.01×10^8	9.35×10^7	9.80×10^7
Jiangsu	1.79×10^8	1.88×10^8	2.08×10^8	2.34×10^8	2.40×10^8	2.47×10^8	2.48×10^8	2.60×10^8
Zhejiang	2.25×10^8	2.27×10^8	2.26×10^8	2.41×10^8	2.42×10^8	2.45×10^8	2.47×10^8	2.53×10^8
Fujian	1.18×10^8	1.29×10^8	1.35×10^8	1.43×10^8	1.43×10^8	1.43×10^8	1.59×10^8	1.58×10^8
Shandong	3.46×10^8	3.56×10^8	3.90×10^8	4.09×10^8	4.29×10^8	4.21×10^8	4.48×10^8	4.77×10^8
Guangdong	2.06×10^8	2.25×10^8	2.35×10^8	2.50×10^8	2.51×10^8	2.50×10^8	2.52×10^8	2.55×10^8
Guangxi	5.84×10^7	6.58×10^7	6.95×10^7	8.33×10^7	9.12×10^7	8.86×10^7	9.04×10^7	8.87×10^7
Hainan	5.15×10^7	5.57×10^7	5.62×10^7	6.22×10^7	6.49×10^7	6.27×10^7	6.87×10^7	7.19×10^7

stabilization. The marine ecological footprint of Shanghai reached its peak in 2013, and then showed a downward trend. The marine ecological footprints of Guangdong and Guangxi provinces were similar, rising year by year up to 2012 and then becoming stable. Except for Shanghai, the reason for the final stabilization of marine ecological footprints in these provinces was the stabilization or decline in their energy ecological footprints, while the decline in the marine ecological footprint in Shanghai was mainly related to the decline of its fishery ecological footprint.

6.4 Decoupling of marine resources and economic development

6.4.1 Decoupling model construction

The "decoupling" theory (i.e., the end of the relationship among interrelated physical quantities) originated from the concept of physics, and has been used by foreign economists to study the relationship between economic development and resource consumption pressure. The Organization for Economic Cooperation and Development's (OECD) decoupling model (OECD, 2002) was first proposed to analyze the relationship between economic development and resource consumption. The model uses the ratio of environmental resource pressure to economic growth as the index, and divides decoupling into absolute decoupling and relative decoupling. Absolute decoupling means that under the scenario of rapid economic growth, resource consumption not only does not increase, but may also begin to decline. Relative decoupling means that the pressure of resource consumption is gradually increasing in the process of economic growth, but the growth rate of resource consumption is smaller than that of economic growth. As the OECD data selection is based only on a base period and end period values, there are unavoidable errors. Thus the Tapio (2005) decoupling model based on elasticity is selected for this study. Subsequently, variable transformation is carried out to construct a decoupling model between marine economic growth and marine ecological footprint growth. Specifically,

$$t_{Ef,O} = \frac{\Delta Ef / Ef}{\Delta O / O} \tag{6.1}$$

where Ef represents the marine ecological footprint (hectares); O stands for marine GDP (CNY10000); and $t_{Ef,O}$ denotes the decoupling elasticity between marine economic growth and the marine ecological footprint. According to Tapio's research and the calculation results of formula (6.1), the decoupling types are divided into three categories: decoupling, negative decoupling, and connection, with eight sub-categories (Onat, Kucukvar, & Tatari, 2014; Wang, 2013) such as strong decoupling and strong negative decoupling. Table 6.3 shows the specific structural status and the assessment basis.

To analyze further the factors affecting the growth of the marine ecological footprint and the elasticity of marine economic growth, and accurately estimate the contribution of each factor to the elasticity, this study uses the exponential decomposition method to decompose the marine ecological footprint. The exponential decomposition method is often applied to the analysis of factors affecting a change in resource consumption. The LMDI (Logarithmic Mean Divisia Index) method of exponential decomposition (Han, Zhong, Yu, Xi, & Liu, 2018; Mathy, Menanteau, & Criqui, 2018; Moutinho, Madaleno, Inglesilotz., & Dogan, 2018) is widely used in scientific research because of its wide–ranging application and no generation of residual value. This chapter uses this method to analyze the factors influencing the growth elasticity of the marine ecological footprint. Considering the interpretation of decomposition results, the additive decomposition model based on LMDI is used. First, the marine ecological footprint can be decomposed as given below:

$$Ef = \sum Ef_i \quad = \sum_i \frac{Ef_i}{Ef} \times \frac{Ef}{O} \times \frac{O}{GDP} \times \frac{GDP}{P} \times P \qquad (6.2)$$
$$= \sum S \times T \times I \times Y \times P$$

where Ef_i represents the ecological footprint (hectare) in the ith type in the composition of the marine ecological footprint. GDP denotes the gross local

Table 6.3 Decoupling states of economic resources.

$\triangle$ Ef	$\triangle$ O	$t_{Ef,O}$	Decoupling states	
<0	<0	$t>1.2$	Decoupling	Decoupling
>0	>0	$0<t<0.8$	Weak decoupling	
<0	>0	$t<0$	Strong decoupling	
>0	>0	$t>1.2$	Expansive negative decoupling	Negative decoupling
<0	<0	$0<t<0.8$	Weakly negative decoupling	
>0	<0	$t<0$	Strong negative decoupling	
>0	>0	$0.8<t<1.2$	Expansive Connection	Connection
<0	<0	$0.8<t<1.2$	Declining connection	

product and P denotes the total local population. $S = \frac{Ef_i}{Ef}$ indicates the proportion of the ecological footprint in the ith type in proportion to the marine ecological footprint, and is used to measure the structural effect of the marine ecological footprint. $T = \frac{Ef}{O}$ indicates the marine ecological footprint (hectares/CNY10000) needed per unit of GDP, and is used to measure technological effects. $I = \frac{O}{GDP}$ refers to the proportion of the marine industry to GDP and is used to measure the industrial structure effect. $Y = \frac{GDP}{P}$ refers to the per capita output of the region (CNY10000/10,000 persons), and is used to measure the output effect. P is used to measure the population effect.

To further decompose formula (6.2) according to the LMDI method, the total variation ($\triangle Ef$) of the marine ecological footprint from time t to $t + 1$ can be expressed as follows:

$$\Delta Ef = Ef_{t+1} - Ef_t = \Delta E_S + \Delta E_T + \Delta E_I + \Delta E_Y + \Delta E_P \qquad (6.3)$$

where ΔE_S, ΔE_T, ΔE_I, ΔE_Y, and ΔE_P indicate the structural effect, technological effect, industrial structure effect, the output effect, and the population effect, respectively, which influence the marine ecological footprint. Referring to past research (Ang, Zhang, & Choi, 1998), the items on the right side of formula (6.3) can be expressed as follows:

$$\Delta E_S = \sum M_{t+1,t} \times \ln \frac{S_{t+1}}{S_t}, \Delta E_T = \sum M_{t+1,t} \times \ln \frac{T_{t+1}}{T_t},$$

$$\Delta E_I = \sum M_{t+1,t} \times \ln \frac{I_{t+1}}{I_t},$$

$$\Delta E_Y = \sum M_{t+1,t} \times \ln \frac{Y_{t+1}}{Y_t}, \Delta E_P = \sum M_{t+1,t} \times \ln \frac{P_{t+1}}{P_t} \qquad (6.4)$$

While,

$$M_{t+1,t} = \frac{Ef_{t+1} - Ef_t}{\ln(Ef_{t+1}/Ef_t)} \qquad (6.5)$$

Therefore,

$$t_{Ef,O} = \frac{\Delta Ef/Ef}{\Delta O/O} = \frac{\Delta E_S/Ef}{\Delta O/O} + \frac{\Delta E_T/Ef}{\Delta O/O} + \frac{\Delta E_I/Ef}{\Delta O/O}$$

$$+ \frac{\Delta E_Y/Ef}{\Delta O/O} + \frac{\Delta E_P/Ef}{\Delta O/O} \qquad (6.6)$$

$$= t_S + t_T + t_I + t_Y + t_P$$

where t_S, t_T, t_I, t_Y, and t_P, represent corresponding decoupling elasticity indices. Thus the decoupling elasticity index $t_{EF, \, O}$ of the marine economy and the ecological footprint can be decomposed into structural decoupling elasticity t_S, technical decoupling elasticity t_T, industrial structure decoupling elasticity t_I, output decoupling elasticity t_Y, and population decoupling elasticity t_P.

The indices presented in this chapter include the regional gross marine product of the coastal areas, the output of fishery resources, the output of marine crude oil, the output of marine natural gas, the output of marine raw salt, the output of mudflat biomass, the output of marine chemical products, the gross marine product of coastal areas, and the population in the coastal areas. Among them, 2008 is chosen as the base period for price deflation of regional gross marine product. The unit for the output of all resources is 10,000 tons. Relevant data come from the China Statistical Yearbook, the China Ocean Statistical Yearbook, the China Fishery Statistical Yearbook, and local marine statistics bureaus.

6.4.2 Results analysis

Through the decoupling evaluation model, we get the rate of change for the marine ecological footprint and for the marine GDP, as well as the decoupling index, the corresponding relationship between marine economic growth and the utilization of marine resources (Table 6.4), and a trend chart for marine GDP—marine ecological footprint variation (Fig. 6.2) from 2008 to 2015. We can see that the growth rates in the marine ecological footprints fluctuated within 0.08 in almost all years from 2009 to 2015, while the marine GDP rate of change basically remained in the range of 0.07—0.17. This shows that while the marine economies of the coastal areas grew, their marine ecological footprints

Table 6.4 Relationship of marine GDP—marine ecological footprint in coastal areas from 2008 to 2015.

Year	ΔEf/Ef	ΔO/O	$t_{Ef,O}$	Decoupling state
2008—09	0.047	0.108	0.433	Weak decoupling
2009—10	0.060	0.171	0.352	Weak decoupling
2010—11	0.080	0.089	0.900	Expansive Connection
2011—12	0.0263	0.103	0.256	Weak decoupling
2012—13	-0.004	0.079	-0.047	Strong decoupling
2013—14	0.017	0.120	0.143	Weak decoupling
2014—15	0.025	0.095	0.264	Weak decoupling

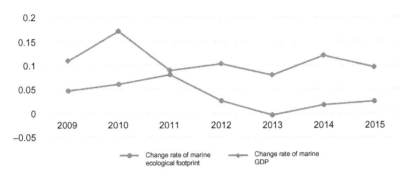

Figure 6.2 Variations in marine GDP–marine ecological footprint change in coastal areas from 2008 to 2015.

grew at a relatively low rate. The relationship between them aligns with the decoupling and coordination conditions between marine resource utilization and marine economy growth. From 2009 to 2015, marine resource utilization and marine economy growth was tended to be stable, namely, weak decoupling. The consumption of marine resources and the marine economy growth showed a positive correlation. The decoupling state in 2010 reflects an expansive connection, while 2012–13 shows strong decoupling. In addition, Fig. 6.2 shows that before 2011, the growth rate in the marine economy in coastal areas had a general trend of fluctuating decline, while the growth rate of the marine ecological footprint showed a fluctuated decline between 2009 and 2011. After 2011, the growth rate of the marine economy in coastal areas slowed and gradually stabilized, and the growth rates of the marine ecological footprint showed a slow downward trend (but always >0). This highlights that the coastal areas should find new economic growth points as soon as possible, improve the speed of marine economic growth, realize the transformation and upgrading of their marine economic structures, and improve the quality and efficiency of marine economic development on the premise of maintaining the slow growth of marine resource utilization.

Next, we evaluated the decoupling state of each coastal area; the results are shown in Table 6.5. Similar to the national situation, the decoupling relationship between the marine economy and the marine ecological footprint in most years between 2008 and 2015 showed a state of weak decoupling. The decoupling states in Zhejiang, Fujian, and Shandong were relatively stable, with weak decoupling in most years and strong decoupling in a few years. This means that there was a stable positive decoupling relationship between the marine economies and

Table 6.5 Evaluation of relationship between marine GDP–marine ecological footprint in coastal areas.

Province	2008–09	2009–10	2010–11	2011–12	2012–13	2013–14	2014–15
Tianjin	Weak decoupling	Weak decoupling	Expansive Connection	Weak decoupling	Weak decoupling	Strong decoupling	Expansive Connection
Hebei	Strong negative decoupling	Weak decoupling	Weak decoupling	Weak decoupling	Strong decoupling	Strong decoupling	Weak decoupling
Liaoning	Weak decoupling	Weak decoupling	Weak decoupling	Strong negative decoupling	Strong decoupling	Weak decoupling	Weak decoupling
Shanghai	Weakly negative decoupling	Weak decoupling	Weak decoupling	Weak decoupling	Expansive Connection	Declining decoupling	Weak decoupling
Jiangsu	Weak decoupling	Weak decoupling	Expansive Connection	Weak decoupling	Expansive Connection	Weak decoupling	Weak decoupling
Zhejiang	Weak decoupling	Strong decoupling	Weak decoupling	Weak decoupling	Weak decoupling	Weak decoupling	Weak decoupling
Fujian	Weak decoupling	Weak decoupling	Weak decoupling	Weak decoupling	Strong decoupling	Weak decoupling	Strong decoupling
Shandong	Weak decoupling	Weak decoupling	Weak decoupling	Weak decoupling	Strong decoupling	Weak decoupling	Weak decoupling
Guangdong	Weak decoupling	Weak decoupling	Expansive Connection	Weak decoupling	Strong decoupling	Weak decoupling	Weak decoupling
Guangxi	Expansive Connection	Weak decoupling	Expansive Connection	Weak decoupling	Strong decoupling	Weak decoupling	Strong decoupling
Hainan	Weak decoupling	Weak decoupling	Expansive Connection	Weak decoupling	Strong decoupling	Expansive Connection	Weak decoupling

marine resource consumption in these areas. Although Tianjin, Jiangsu, Guangdong, Guangxi, and Hainan presented positive decoupling states in most years, there were still expansive connections in some years and the growth in the marine economy still depended on the increase in marine resource consumption. In Hebei, Liaoning, and Shanghai, from 2008 to 2015, due to the unreasonable structure of their marine economies and exploitation of marine resources, the relationship between their marine economies and marine resource consumption showed a negative decoupling state, that is, the consumption of marine resources was rising while the marine economies were declining.

To further explain the decoupling relationship between the marine economies and marine resource consumption in coastal areas, according to formulas (6.2)−(6.6), the LMDI decomposition method was used to analyze the effects of structure, technology, industrial structure, output, and population. The effects of marine resource consumption and marine economic growth were measured and the driving factors of the decoupling state between marine economic growth and resource consumption in coastal areas were explored. The decomposition results of decoupling resource consumption and marine economic growth in coastal areas during 2008 and 2015 are shown in Table 6.6.

Table 6.6 shows that the main factor that inhibited the consumption of marine resources in coastal areas was the technological effect, except in 2010 and 2011. In some years, the industrial effect inhibited the consumption of marine resources, but with weak effect. The main effects that contributed to the consumption of marine resources were the output effect and the population effect. During 2008 and 2011, the growth rates of the marine footprint were 4.7%, 6%, and 8%, respectively, with an average annual growth rate of 6.23%. Although the improvement in the marine resource utilization level reduced the consumption of marine resources to a certain extent, it was far from enough to offset the increase in marine resource consumption caused by various factors such as the population effect. From 2011 to 2015, the growth rate of the marine footprint was 2.63%, −0.04%, 1.7%, and 2.5%, respectively. The annual average growth rate was about 1.7%, which was significantly lower than during 2008−11. By observing the decoupling decomposition results, we can see that the technological effect continued to grow stronger, inhibiting the increase in resource consumption, and reaching a peak between 2012 and 2013. However, the output effect still continued to promote marine resource consumption growth.

Table 6.6 Decomposition of decoupling factors between marine GDP—marine ecological footprint in coastal areas from 2008 to 2015.

Year	t_S	t_T	t_I	t_Y	t_P	$t_{Ef,O}$	Major promoting factors	Major inhibiting factors
2008—09	0	−0.539	−0.0262	0.886	0.112	0.433	Output effect	Technological effect
2009—10	0	−0.598	0.231	0.635	0.0841	0.352	Output effect	Technological effect
2010—11	0	−0.0957	−0.207	1.13	0.0780	0.900	Output effect	Industrial effect
2011—12	0	−0.709	0.0751	0.823	0.0660	0.256	Output effect	Technological effect
2012—13	0	−1.01	−0.140	1.03	0.0751	−0.047	Output effect	Technological effect
2013—14	0	−0.809	0.304	0.599	0.0497	0.143	Output effect	Technological effect
2014—2015	0	−0.704	0.180	0.721	0.0654	0.264	Output effect	Technological effect

To sum, the key factors affecting the decoupling elasticity between marine economic growth and marine resource consumption were the technology effect and the output effect. The effects of structure and population on decoupling elasticity were not obvious.

Thus the decoupling of marine economic growth and marine resource consumption should begin with the technology and industrial structure. Technologies should be continuously innovated to improve the unit utilization efficiency of marine resources. Policies should be enacted to support green and environmental-friendly marine emerging industries and eliminate backward industries with high-resource consumption and low-resource utilization.

6.5 Conclusion and Suggestion

6.5.1 Conclusion

The modified marine ecological footprint model is used to measure the marine ecological footprint in coastal areas and the degree of coordination between marine economic growth and marine resource utilization in coastal areas is evaluated using the decoupling model. Research results show that: (1) From 2008 to 2015, except for several years, the marine ecological footprint of coastal areas showed an overall upward trend and the marine utilization for energy was the main component of the marine ecological footprint. (2) Among the coastal areas, Shandong had the largest marine ecological footprint, with the marine ecological footprints in most coastal areas rising at first and then becoming stable. The marine ecological footprints of some coastal areas increased year by year. (3) From 2008 to 2015, the relationship between the marine economy and the marine ecological footprint in China's coastal areas was mainly in the weak decoupling state. Marine economic growth and marine resource consumption were decoupled to a certain extent. According to the decomposition results on decoupling elasticity between the marine economy and the marine ecological footprint, technological progress was the main reason for the decoupling.

6.5.2 Suggestion

As marine resources are not infinite, marine economic development, inevitably, will be bound by them at a certain stage. To clear this bottleneck, the dependence on the exploitation of marine resources in the process of marine economic development should be reduced gradually to decouple marine resources and marine economic development. With the help of scientific and technological innovation, marine resources can be more effectively utilized and allocated, and the industrial structure can be optimized gradually to change the development mode of the marine economy from being extensive to being intensive.

First, the development and utilization of marine resources should continue to increase in depth and breadth. It is necessary not only to coordinate the development of the marine economy and the development of marine resources at the national level, but also to arrange the development of marine resources according to the specific conditions in coastal areas. Second, the sustainable development of the marine economy and marine resources should be promoted. China needs to foster the transformation of its marine economy, along with new industry, upgrade backward industries, improve backward development modes, actively develop marine culture, and choose the direction for green marine economic development that enables low consumption and pollution. Last, China needs to strengthen marine scientific, technological, and institutional innovation and the cooperation among its coastal areas to realize the complementary advantages of these marine resources in different regions to enhance the supply capacity of marine resources and the sound development of the overall marine economy.

References

Ang, B. W., Zhang, F. Q., & Choi, K. H. (1998). Factorizing changes in energy and environmental indicators through decomposition. *Energy.*, *23*(6), 489−495.

Barange, M., Cheung, W. W. L., Merino, G., & Perry, R. I. (2010). Modeling the potential impacts of climate change and human activities on the sustainability of marine resources. *Current Opinion in Environmental Sustainability.*, *2*(5-6), 326−333.

Figge, F., Young, W., & Barkemeyer, R. (2014). Sufficiency or efficiency to achieve lower resource consumption and emissions? The role of the rebound effect. *Journal of Cleaner Production*, *69*, 216−224.

Fock, H. O., Kloppmann, M. H. F., & Probst, W. N. (2014). An early footprint of fisheries: Changes for a demersal fish assemblage in the German Bight from 1902−1932 to 1991−2009. *Journal of Sea Research*, *85*, 325−335.

Gogoberidze, G. (2012). Tools for comprehensive estimate of coastal region marine economy potential and its use for coastal planning. *Journal of Coastal Conservation*, *16*(3), 251−260.

Guillaumont, P. (2010). Assessing the economic vulnerability of small island developing states and the least developed countries. *Journal of Development Studies*, *46*(5), 828−854.

Halpern, B. S., Longo, C., & Hardy, D. (2012). An index to assess the health and benefits of the global ocean. *Nature.*, *488*(7413), 615−620.

Han, H., Zhong, Z., Yu, G., Xi, F., & Liu, S. (2018). Coupling and decoupling effects of agricultural carbon emissions in China and their driving factors. *Environmental Science & Pollution Research International, 9,* 1−14.

Jones, E. M., Doblin, M. A., Matear, R., & King, E. (2015). Assessing and evaluating the ocean-colour footprint of a regional observing system. *Journal of Marine Systems, 143,* 49−61.

Li, B., Yang, Z., & Su, F. (2016). Vulnerability measurement of Chinese marine economic system based on set pair analysis. *Scientia Geographica Sinica, 36*(01), 47−54.

Managi, S., Opaluch, J. J., Jin, D., & Grigalunas, T. A. (2005). Technological change and petroleum exploration in the Gulf of Mexico. *Energy Policy., 33*(5), 619−632.

Mathy, S., Menanteau, P., & Criqui, P. (2018). After the Paris agreement: Measuring the global decarbonization wedges from national energy scenarios. *Ecologica Economics, 150,* 273−289, Post-Print.

Monfreda, C., Wackernagel, M., & Deumling, D. (2004). Establishing national natural capital accounts based on detailed Ecological Footprint and biological capacity assessments. *Land Use Policy, 21*(3), 231−246.

Moutinho, V., Madaleno, M., Inglesilotz. R., Dogan, E. 2018. Factors affecting CO2 emissions in top countries on renewable energies: A LMDI decomposition application. *Renewable & Sustainable Energy Reviews.*

OECD. (2002). *Indicators to Measure Decoupling of Environment Pressures From Economic Growth.* Paris: OECD.

Onat, N. C., Kucukvar, M., & Tatari, O. (2014). Scope-based carbon foot print analysis of US residential and commercial buildings: An input output hybrid life cycle assessment approach. *Building and Environment, 72,* 53−62.

Rees, W. E. (1992). Ecological footprints and appropriated carrying capacity: What urban economics leaves out. *Environment and Urbanization, 4*(2), 121−130.

Samonte-Tan, G. P., White, A. T., Tercero, M. A., Diviva, J., Tabara, E., & Caballes, C. (2007). Economic valuation of coastal and marine resources: Bohol marine triangle, Philippines. *Coastal Management., 35*(2-3), 319−338.

Side, J., & Jowitt, P. (2002). Technologies and their influence on future UK marine resource development and management. *Marine Policy., 26*(4), 231−241.

Sun, C. Z., Zhang, K. L., Zou, W., & Zeyu, W. (2015). Study on regional system of man-sea relationship and its synergetic development in the coastal regions of China. *Geographical Research., 34*(10), 1824−1838.

Tapio, P. (2005). Towards a theory of decoupling: Degrees of decoupling in the EU and the case of road traffic in Finland between 1970 and 2001. *Transport Policy., 12*(2), 137−151.

Wang, D. L. (2013). Research on the relationship between the change of marine industry structure and the growth of marine economy in Fujian Province. *Ocean Development and Management, 30*(09), 85−90.

Wernberg, T., Thomsen, M. S., & Connell, S. D. (2013). The footprint of continental-scale ocean currents on the biogeography of seaweeds. *Plos One., 8*(11), e80168.

CHAPTER SEVEN

Analysis of coupling among marine resources, environment, and economy in China

7.1 Introduction

In the 21st century, all marine countries, including China, regard the development and utilization of marine resources as an important national development strategy. In this context, after 40 years of rapid development following the reform and opening up, China's gross marine product had exceeded CNY 7 trillion by 2016 and the marine economy had become the main driving force of China's national economic growth. In recent years, with the support of the strategy of accelerating the construction of "ocean power," the vitality of China's marine economy has become progressively more prominent. Large-scale state-owned enterprises have begun to invest worldwide, including not only in traditional marine industries such as fisheries, shipping ports, and marine oil and gas but also in emerging marine industries such as comprehensive utilization of seawater, marine environmental protection, and social services. However, it must be noted that in the process of the rapid development of the marine economy, excessive consumption of marine resources and excessive discharge of waste have caused a great burden on the marine ecological environment, resulting in the gradual highlighting of a contradiction between the marine economy, resources, and environment. In addition, population, economic scale, and spatial distribution of coastal areas will continue to expand with the continuous development of China's coastal economic belt. The pressure and demand of social and economic activities on marine resources and environment will continue to increase, resulting in resource and environmental problems that will increasingly affect the sustainable and healthy development of the marine economy. In this context, it is of great practical significance to study the coupling among marine resources, environment, and economy.

Sustainable Marine Resource Utilization in China.
DOI: https://doi.org/10.1016/B978-0-12-819911-4.00007-2

7.2 Literature review

The purpose of the coordinated development of regional resources, environment, and economy is to achieve sustainable development in the region, which is conducive to a virtuous circle of social development through the interaction and cooperation among the three subsystems and the elements within each. Therefore finding the balance point between economic development, resource allocation, and the ecological environment in coastal areas by scientific methods is conducive to the coordinated development of coastal provinces. Grossman and Krueger (1991), the first scholars to study the relationship between economy and environment, used the Environmental Kuznets Curve to explore the relationship between environmental deterioration and income level. Since then, many scholars have begun to use this theory to study the relationship between economic development and environmental pollution. Environmental pollution indicators are mainly expressed by suspended particulate matter, nitrogen oxide emissions, and sulfur dioxide or carbon dioxide emissions (Orubu & Omotor, 2011; Stern & Common, 2001; Maddison, 2006; Saboori, Sulaiman, & Mohd, 2012). However, these studies mainly focused on the effect of a single environmental pollution index on economic development, which cannot express the overall situation of the ecological environment. Therefore in order to reflect the overall development level of the ecological environment, many scholars have carried out a comprehensive evaluation of the ecological environment and the level of economic development. For example, Liem et al. (2002) and Li, Min, and Tan (2004) used analytic hierarchy process and a multilevel fuzzy comprehensive evaluation model to evaluate the regional ecological environment. Wang, Liu, and Yang (2008) and Chen, López, and Walker (2014) put forward a vulnerability assessment and analyzed the vulnerability of the ecosystem and its changes. Read and Field (2003) and Field (2003) both studied and analyzed the development and management of marine resources based on the marine environment. Jonathan and Paul (2002), by predicting the future trend of development and management of marine resources in Britain, believed that the development of renewable marine energy would be the main driving factor of marine technological progress. Managi, Opaluch, Jin, and Grigalunas (2005) studied the effects of oil and gas exploration in the Gulf of Mexico from 1947 to 1998 by establishing an index variable of technological change and predicted the changing trend of marine resource utilization rate with the growth of the marine economy. Gogoberidze (2012) analyzed the real potential of the marine economy in coastal areas for strategic and spatial planning of coastal development. Samonte, White, Tercero, and John (2007) introduced the net benefits of natural resources in the marine triangle of Baohe Island, Philippines, to measure the important life support functions of coastal and marine ecosystems, and emphasized the importance of marine resource management. Halpern, Longo, Hardy, Mcleod, and Samhouri (2012) made an empirical analysis of the health status of the human—ocean coupling system from the perspective of sustainable management of the marine economic system. Manuel, William, Merino, and Perry (2010) explored the synergistic effects of climate change and human activities on marine ecosystems through coupling modeling. Reuveny (2002) analyzed the relationship among economic growth, scarcity of resources, and growing conflicts in underdeveloped countries and believed that balancing economic growth and coordinating economic contraction in developing countries could lead to overall economic growth. Grimaud and Rouge (2008) introduced nonrenewable and nonpolluting labor resources into the general equilibrium framework of endogenous growth and studied the effects of resource constraints on the economy.

In addition, coupling theory is often applied to the study of the relationship between the economy and environment by scholars (Vefie, 1996; Diao, Zeng, & Tam, 2009; Salvador, José, & Mariana, 2007). For example, Di, Porter, and Tracey (2003) combined the input—output model of the coastal economy with the marine food web model to establish an economy—ecology model. Zhang, Su, Li, and Sang (2008) used the methods of principal component and regression analysis to establish a coordinated evaluation method to study the interaction of population, economy, space, and environment in the urban scale. Yasuhiro and Kazuhiro (2005) studied how urbanization and population transformation are related to each other through the advantages and disadvantages of population concentration and explored the adverse effects of the ecological environment on the process of urbanization. With the deepening of the sustainable development concept there has also been wide concern about the relationships among economic development, resource endowment, and ecological environment. Slesser and Hounam (1984) systematically studied the relationship between regional resources and population growth, economic development, social progress, and environmental pollution, starting from the interaction between economy, resources, and environment. Allan, Hanley, Mcgregor, and Turner (2007) studied and analyzed the relationship between economic development, resources, and environment from the perspective of input and output. Oliverira and Antunes (2011) established a multisectoral economic—energy—environment model to conduct a forward-looking analysis of changes in economic structure and energy systems. Tiba and Omri (2017) used data from 1978 to 2014 to study the dynamic relationship among economy, resources, and environment in the sample areas.

Because resource endowment plays an important role in the economic development and ecological environment system of a country or region, the degree of coordinated development among them is related to the level of sustainable economic development. At present, most of the research in this field is carried out from the national or provincial point of view, while the research on a specific industry, especially the marine industry, is relatively rare. In addition, through the study of existing literature on system coupling, we find that different scholars have different ways of calculating the coupling degree, and some of the literature has a problem with misjudging the range of coupling degree. Therefore on the basis of calculating the coupling degree of the marine economic—resource—environment system in 11 coastal provinces of China from 2006 to 2015, this study revises the coupling degree formula. We then use relevant data from 2006, 2009, 2012, and 2015 to reflect the changing process of the coupling degree of the marine economy—resource—environment system over the research period.

7.3 Coupling index system construction and calculation model

7.3.1 Index system construction and analysis

This study screens 17 relevant indexes and results from three levels, marine resources, environment, and economy are shown in Table 7.1. Among them, mariculture production, marine petroleum resources production, marine salt production, per capita coastline length, per capita sea area, and other indexes are selected for the marine resource system; quantity of up-to-standard industrial wastewater in coastal areas, comprehensive utilization of industrial solid waste in coastal areas, number of coastal

Table 7.1 Comprehensive evaluation indexes of marine resource endowment, eco-environment, and economic development system.

System	Evaluation index
Marine resources	Mariculture production
	Marine petroleum resources production
	Marine salt production
	Per capita coastline length
	Per capita sea area
Ecological environment	Quantity of up-to-standard industrial wastewater in coastal areas
	Comprehensive utilization of industrial solid waste in coastal areas
	Number of coastal pollution control projects
	Total amount of industrial wastewater discharged directly into the sea
	Discharge amount of industrial solid waste
	Proportions of the first and second types of water quality
Economic development	Proportion of marine gross product to GDP
	Proportion of marine secondary industry production
	Proportion of marine tertiary industry production
	Marine economy capital stock
	Proportion of marine technological staff
	Number of patent applications for scientific research in marine scientific research institutions

pollution control projects, total amount of industrial wastewater discharged directly into the sea, discharge amount of industrial solid waste, and the proportions of the first and second types of water quality are selected for the marine environment; proportion of marine gross product to GDP, proportion of marine secondary industry production, proportion of marine tertiary industry production, marine economy capital stock, proportion of marine technological staff, and number of patent applications for scientific research in marine scientific research institutions are selected for the marine economy. All data come from each year's *China Ocean Statistical Yearbook*.

The marine resource system includes mariculture and the development volume of related resources, which measure the capacity of resource development and utilization. Related marine resources include marine oil, sea salt, coastline utilization, and confirmed sea area indexes. In addition, the length of per capita coastline reflects the degree of per capita possession of coastline resources, while the per capita sea area reflects the per capita possession of the confirmed sea area.

The marine environmental system includes the proportion of the first and second types of water quality (marine water resource is an important part of the marine ecological environment system, reflecting the degree of marine environmental protection in the region), the quantity of up-to-standard industrial wastewater in coastal areas (referring to the amount of industrial wastewater discharged from the coastal areas that reaches the national or local discharge standards), the amount of comprehensive utilization of industrial solid waste in coastal areas, number of coastal pollution control projects, discharge amount of industrial solid waste, and total amount of industrial wastewater discharged directly into the sea (meaning that wastewater is discharged directly into the sea through the sewage draining exit of a factory). The higher the amounts of industrial solid waste and industrial wastewater discharged directly into the sea, the greater the damage.

The marine economic system is the most direct factor affecting the sustainable development of the marine industry and marine ecological resources and environment. The moderate development of the marine economy is more conducive to social progress. Because the gross product of the marine industry reflects the total economic benefit of a region through the exploitation and utilization of marine resources, this study uses the proportion of gross product of marine industry in a region to its GDP to reflect the contribution of marine industry development to the total economic development in the region. The marine industrial structure of coastal areas in China is represented by the proportions of the second and tertiary marine industry production. The stock of marine economic capital refers to the total amount of marine capital in each province every year, which is the sum of capitals invested in the marine economy. Usually, it measures the existing scale and level of the marine economy. The proportion of marine scientists and technicians represents the talent reserve that supports the development of the marine economy (Qin, Sun, & Wang, 2014; Sun & Wang, 2012). The number of patent applications for scientific research in marine scientific research institutions reflects the level of scientific and technological innovation in the marine economy.

7.3.2 Data standardization

As the data of each coastal provincial subsystem are dimensionalized, it is impossible to compare them directly. Therefore we should first render the

evaluation indexes dimensionless. This chapter uses the deviation standardization method to do the following processing:

When the index is positive:

$$y_{ij} = \frac{x_{ij} - \min(x_j)}{\max(x_j) - \min(x_j)} \tag{7.1}$$

When the index is negative:

$$y_{ij} = \frac{\max(x_j) - x_{ij}}{\max(x_j) - \min(x_j)} \tag{7.2}$$

In these formulae, x_{ij} represents the original value of index j of area i;
$\min(x_j)$ and $\max(x_j)$ represent the minimum and maximum values of
index j in all regions, respectively.

7.3.3 Entropy method for weighting

After data standardization, we use the comprehensive development scores
of marine resources, environment, and economy to evaluate the development status of each system. Before calculating the comprehensive evaluation score of each system, the weight of each index in the system should
be calculated first. Weight calculation methods are divided into subjective
and objective methods. Considering the fuzziness and uncertainty of the
index selection, this study calculates the weight coefficients of each index
by using the objective entropy method to ensure the objectivity of the
results. This avoids the deviation caused by subjective factors to a certain
extent and realizes the more reasonable weighting of the factors affecting
the marine ecological carrying capacity. That is, it uses the inherent information of evaluation indexes to judge the utility value of the index.
Then, the comprehensive evaluation scores of resource endowment, ecological environment, and economic development system of 11 coastal
provinces from 2006 to 2015 are calculated. The calculation method of
the comprehensive evaluation score is as follows:

$$U_{k,i} = \sum_{j=1}^{m} \lambda_{ij} x'_{ij}, \quad \sum_{j=1}^{m} \lambda_{ij} = 1 \tag{7.3}$$

In the formula, $U_{k,i}$ refers to the comprehensive evaluation score of
the marine resource, environment, and economy subsystem in region i
and λ_{ij} is the weight of the index in each subsystem.

7.3.4 Coupling model

The key of the system from disorder to order lies in the interaction of system order parameters. The coupling degree is an index to measure the degree of interaction between systems or elements. Many scholars have discussed how to calculate the system coupling degree. However, there are some errors in judging the range of the coupling degree formula.

Based on the concept of capacity coupling and the model of the capacity coupling coefficient in physics, and with reference to the calculation model of the multisystem coupling degree, this study establishes the calculation model of a three-system coupling degree. In the model, C_{ere} is the C system coupling degree; U_{econ}, U_{res}, and U_{env} are comprehensive scores of marine economy, resources, and environmental systems, respectively.

$$\text{Cere} = \left[\frac{U_{econ} \times U_{res} \times U_{env}}{(U_{econ} + U_{res}) \times (U_{econ} + U_{env}) \times (U_{res} + U_{env})} \right]^{1/3} \qquad (7.4)$$

Then, the range of values of C_{ere} is calculated and analyzed. Because $U_{econ} + U_{res} \geq 2\sqrt{U_{econ} \times U_{res}}$, $U_{econ} + U_{env} \geq 2\sqrt{U_{econ} \times U_{env}}$, and $U_{res} + U_{env} \geq 2\sqrt{U_{res} \times U_{env}}$, if, and only if, the comprehensive scores of the three systems are equal in pairs can the equation hold. The above three inequalities can be multiplied to obtain:

$$(U_{econ} + U_{res}) \times (U_{econ} + U_{env}) \times (U_{res} + U_{env}) \geq 8 U_{econ} \times U_{res} \times U_{env} \qquad (7.5)$$

From formulas (7.4) and (7.5), we can obtain

$$C_{ere} = \left[\frac{U_{econ} \times U_{res} \times U_{env}}{(U_{econ} + U_{res}) \times (U_{econ} + U_{env}) + (U_{res} + U_{env})} \right]^{1/3} \leq \left(\frac{1}{8} \right)^{1/3} = \frac{1}{2} \qquad (7.6)$$

Thus the range of coupling degree of the three systems calculated by formula (7.3) is [0, 1/2]. Similarly, when formula (7.4) is used to calculate the coupling degree of two systems, the range of the coupling degree is also [0, 1/2]. Some scholars have used the formula (7.4) in calculating the coupling degree of two systems. However, because the maximum value of the coupling degree calculated by the formula (7.4) is not more than 1/2, there may be a shortcoming in that the coupling degree is underestimated.

In addition to formula (7.4), some scholars have used the following formula in calculating the coupling degree of three systems:

$$C_3 = \left[\frac{U_1 \times U_2 \times U_3}{\left((U_1 + U_2 + U_3)/3\right)^3} \right]^{1/3} \tag{7.7}$$

In the formula, C_3 is the system coupling degree; U_1, U_2 and U_3 are comprehensive scores of marine economy, resource, and environmental systems, respectively. The results of the mean in equation prove that

$$\frac{U_1 + U_2 + \cdots + U_n}{n} \geq \sqrt[n]{U_1 \times U_2 \times \cdots \times U_n} \tag{7.8}$$

From formula (7.8), we can know that the formula holds if, and only if, $U_1 = U_2 = \cdots = U_n$, at which time formula (7.7) obtains the maximum value of 1.

Therefore based on formula (7.7) this chapter calculates the coupling degree of the marine economy—resource—environment system and then uses the objective quartile method to divide the coupling level into four categories: low, medium, medium high, and high levels. Compared with the subjective method of threshold setting adopted by many scholars, the quartile method is more objective in dividing the coupling types and helps to divide the 11 coastal provinces of China into 4 categories according to the values of the coupling degree.

7.3.5 Variable coefficient model

This chapter calculates the spatial differentiation status of the coupling degree of marine resource, environmental, and economic systems by using a variable coefficient. Formulae are as follows:

$$\sigma = \sqrt{\frac{1}{N} \sum_{i=1}^{N} (x_i - v)^2} \tag{7.9}$$

$$CV = \left(\frac{\sigma}{v}\right) \times 100\% \tag{7.10}$$

In the formulae, CV is the variable coefficient of the system coupling degree, σ represents the standard deviation of the system coupling degree, v indicates the average value of the system coupling degree, x_i is the coupling degree value of the first coastal province, and N is the number of coastal provinces.

7.4 Empirical analysis of coupling

7.4.1 Coupling measurement and analysis

The evaluation index system of the marine resource, environmental, and economic system is established based on the above. The weight of each index is calculated through the entropy method. On this basis, the comprehensive evaluation scores of each subsystem are calculated. Finally, the coupling degrees between subsystems are calculated with consideration of the coupling degree calculation formulae. According to the values of coupling degree, coastal areas are divided into three types: the northern, eastern, and southern coastal regions. The results are shown in Table 7.2, according to which, the overall coupling effect of all coastal areas is analyzed and Fig. 7.1 is obtained.

It can be seen from Fig. 7.1 that the overall development trend of China's marine resources—environment—economy system has improved in recent years. The system coupling degree of Guangxi has risen year by year, from 0.6911 to 0.9168, gradually realizing the development trend from a low-to-high coupling degree. However, though the coupling degree of Guangxi had been increasing rapidly, it was still at a low level amongst coastal provinces. The coupling degree of Shanghai had been in a downward state until 2014 and 2015 and is still at the lowest level, indicating that the level of marine economic development in Shanghai was extremely disharmonious with marine resources and environment, and the interaction ability was poor. The coupling degree of Tianjin was in a changing trend of 0.9949—0.8748—0.9041 and in a significant fluctuating state, changing from the high level in 2006 to the low-level state in 2010 and then increasing in 2015. The system coupling degrees of Hebei, Jiangsu, and Shandong provinces all showed a trend of rising—falling—rising, and the amplitudes in the first year were high. Although the coupling degree of Liaoning province had always been in a fluctuating situation, it stably ranked in the fourth or fifth place, indicating that although the coordinated development of marine resources, environment, and economic system in Liaoning province had been at a medium high level during the research period, its progress was relatively slow and there was no significant achievement. Guangdong province and Zhejiang province were basically in an upward trend. The coupling degree of Guangdong province had gradually developed from a medium high to high level

Table 7.2 Coupling degrees of coastal areas from 2006 to 2015.

		2006	2007	2008	2009	2010	2011	2012	2013	2014	2015	Mean value
North	Tianjin	0.947	0.945	0.959	0.976	0.933	0.942	0.959	0.937	0.941	0.942	0.948
	Hebei	0.800	0.897	0.906	0.872	0.867	0.873	0.835	0.830	0.892	0.882	0.865
	Liaoning	0.939	0.967	0.969	1.000	0.951	0.985	0.990	0.987	0.989	0.966	0.975
	Shandong	0.923	0.976	0.978	0.977	0.992	0.991	0.998	0.997	0.990	0.994	0.981
East	Shanghai	0.748	0.758	0.727	0.696	0.608	0.600	0.579	0.555	0.553	0.573	0.640
	Jiangsu	0.774	0.848	0.801	0.821	0.842	0.815	0.848	0.854	0.840	0.805	0.825
	Zhejiang	0.746	0.721	0.764	0.779	0.783	0.772	0.802	0.799	0.787	0.787	0.774
South	Fujian	0.873	0.925	0.916	0.903	0.857	0.917	0.933	0.948	0.945	0.943	0.916
	Guangdong	0.848	0.882	0.936	0.940	0.958	0.946	0.968	0.970	0.960	0.941	0.935
	Guangxi	0.696	0.684	0.698	0.728	0.703	0.762	0.740	0.771	0.814	0.805	0.740
	Hainan	0.925	0.922	0.924	0.920	0.918	0.917	0.922	0.929	0.908	0.915	0.920

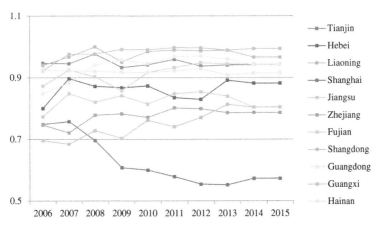

Figure 7.1 Changing trends of coupling degrees of marine resources, environment, and economy.

while that of Zhejiang province had been in an upward state from 2006 to 2010, declining slightly after 2010 and gradually increasing to 0.9024 after falling to 0.8875 in 2011. However, compared with other provinces in the same year, the coupling degree of Zhejiang was still at a low level. Fujian province as a whole was in the state of rising—falling—rising—falling. Its coupling degree increased from 0.8644 to 0.9785 during 2006 and 2007, then declined until 2011, gradually rose to 0.9987 in 2013, and decreased again. Hainan province has been in a state of high coupling all the time.

7.4.2 Coupling degree spatial differentiation patterns

Fig. 7.2 shows the spatial distribution of the coupling degree values of the 11 coastal provinces in China in 2006, 2009, 2012, and 2015, respectively. From the comparison of these four years we can see that the divisions of coupling degrees of the three marine systems in provinces other than Shandong had all changed to some extent.

In 2006 Tianjin and Hainan fell into the high level, which indicates that the internal factors of marine resources, environment, and economy in these two areas had improved mutually and reached a benign resonance coupling. However, over time, the system coupling degree of Tianjin declined from the high coupling level in 2006 to the medium high in 2009. After the medium level in 2012 Tianjin was in the low coupling level until 2015. This shows that the imbalance of the development of Tianjin's marine system has been increasing year by year, and the pressure

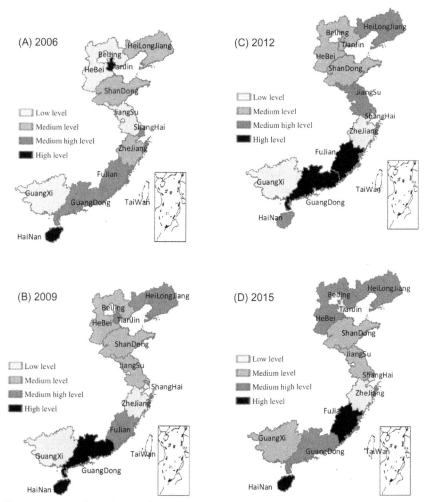

Figure 7.2 Coupling degree distribution in 11 coastal provinces in 2006, 2009, 2012, and 2015.

on marine resources and environmental systems has also been increasing. The damage to the marine system caused by economic development has exceeded the ocean's ability to withstand, thus creating a situation of low coupling. However, because of the relatively low level of economic development and the relatively small amount of industrial pollutants discharged into sea, the system coupling of Hainan province fell into the high level in 2006, 2009, and 2015, with the exception of 2012.

Besides, the system coupling degrees of marine resources, environment, and economy in Hebei, Liaoning, Shanghai, Jiangsu, Zhejiang,
Fujian, Guangdong, and Guangxi also changed. Liaoning province's system coupling degree had gradually developed from the medium level in
2006 to a medium high level in 2009 and then remained unchanged in
2012 and 2015. Hebei province had gradually developed from the low
level in 2006 to the medium level in 2009 and further developed to the
medium high level in 2015. The system coupling degree of Jiangsu province underwent a trend of low—medium—medium high—medium level.
However, the coupling degree of subsystems cannot explain the development status of each subsystem itself. For example, although the system
coupling degree in Hebei province increased, the comprehensive scores of
the economic, resource, and environmental subsystems changed from
0.1744, 0.0903, and 0.4614 in 2006 to 0.1875, 0.0912, and 0.3219 in
2015, respectively. This indicates that the biggest reason for the increase
in the coupling degree in Hebei province was the decline of the comprehensive score of the marine environmental system. The system coupling
degree of Shanghai was at a high level in 2006, while at a low level in all
other periods. Similarly, the system coupling degree of Zhejiang province
was at a medium level in 2006 while remaining at a low level since then.
The system coupling degree of Guangxi Zhuang Autonomous Region
increased in 2015 while remaining at a low level in all other times. This
shows that the development of the marine economy, resources, and environment in these areas was unbalanced and their mutual promotion was
poor. The development of the marine economy—resources—environment
system in Shanghai was imbalanced with a low coupling degree because
of rapid economic development, lower marine resource content than
other coastal provinces, and greater damage to the marine ecological environment. The system coupling degrees of Fujian province and
Guangdong province were at a medium high (high) level at all times,
meaning that there was harmony among each system or each element
within each system.

7.4.3 Evolution of coupling degree variable coefficient

In this study, 11 coastal areas are divided into northern (including Tianjin,
Shandong, Hebei, and Liaoning), eastern (including Shanghai, Jiangsu,
and Zhejiang), and southern (including Fujian, Guangdong, Guangxi, and
Hainan) coastal areas. According to formulae (8—16) and (8—17), variable

coefficients of the overall coupling degree of coastal provinces in 2006, 2009, 2012, and 2015, as well as those of northern, eastern, and southern coastal areas, were calculated to analyze the evolution of the coupling degree variable coefficients and the differences among them.

As can be seen from Fig. 7.3, although the coupling degree variable coefficient of the marine resources—environment—economy system in coastal provinces declined from 0.1297 in 2006 to 0.0645 in 2009, then increased to 0.0956 in 2012, and remained unchanged in 2015, it was generally in a downward trend from 0.1297 in 2006 to 0.0936 in 2015. The changing trend of the coupling degree variable coefficient in the northern coastal province was similar to that of its overall coupling degree variable coefficient, that is, decline—increase—decline. The coupling degree variable coefficient in the southern region generally showed a downward trend from 0.1239 in 2006 to 0.0324 in 2015, indicating that fluctuations of overall coupling degree and coupling degrees of northern and southern coastal areas were in a declining trend and synergistic development had been gradually realized. On the contrary, the coupling degree variable coefficient in the eastern region showed an upward trend, rising from 0.0773 in 2006 to 0.1469 in 2012. Although it declined in 2015 it still reached 0.1378. This shows that the development synergy of Shanghai, Jiangsu, and Zhejiang in the eastern coastal areas was poor.

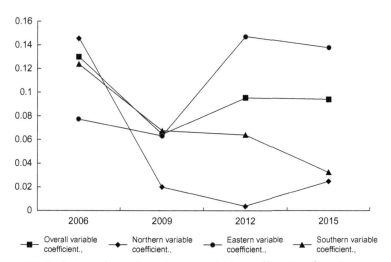

Figure 7.3 Evolution of coupling degree variable coefficients of marine resource, environmental, and economic systems in coastal provinces.

Horizontal comparison shows that fluctuations of coupling degrees were significant during 2006 and 2009. In 2006 ranking of fluctuations of coupling degree from large to small was northern, southern, and eastern coastal areas, meaning that the coupling degree of marine resources—environment—economy system in eastern coastal provinces was less discrete and spatial differences among each province were small. In 2009 as the coupling degree of each province in the northern region was concentrated at a medium high or medium level and the development gap between different regions was reduced, the coupling degree variable coefficient in the northern coastal region decreased significantly. However, with rapid economic development in the eastern coastal region, the coupling degree variable coefficient of the eastern coastal region increased rapidly in 2012, which exceeded the northern and southern coastal areas. Until 2015 the diversity coefficient was still the largest in the eastern coastal region, followed by that in the southern and then the northern coastal regions.

7.5 Conclusion and suggestions

7.5.1 Conclusion

Based on the analysis of the coupling of marine resources, environment, and economy in the coastal areas of China, the following conclusions are drawn:

1. From 2006 to 2015 coupling degrees of coastal provinces in China fluctuated obviously with significant spatial differences, but the overall level of the coupling degree was in a rising trend. Specifically, because of the rapid economic development and comparative scarcity of marine resource endowment in Shanghai and Tianjin, the coupling degree of Shanghai's marine resources—environment—economy system was always in a downward trend and that of Tianjin also declined in fluctuations. Coupling degrees in other regions had certain fluctuations, and most provinces showed an upward trend, indicating that the marine economy had a certain consistency with the changes in the marine resources and environment in recent years.

2. According to the spatial differentiation of coupling degrees, except Shandong province, coupling degree divisions of the marine

resource—environment—economy systems in other provinces have all changed to a certain extent, and all coupling degrees except those in Tianjin and Zhejiang had developed toward the high level. In addition, differences in coupling degrees among eastern coastal provinces were significant and in a rising trend, contrary to the situation in the northern and southern regions. This indicates that synergistic development of the marine resource—environment—economy system among each province in the northern and southern regions was improving.

7.5.2 Suggestions

Against the background of increasing attention to marine resources, we should also focus on the ocean when accelerating economic growth and also accelerate transformation and upgrading. Based on previous empirical studies, the following policy recommendations and enlightenment are drawn.

1. The coupling degree of marine economy—resource—environment in China's coastal provinces has been in an obvious upward trend. While accelerating the economic development of coastal areas, we should pay attention to the protection of marine resources and coastal environment, seek cooperation in all aspects in coastal areas, and jointly tackle marine environmental problems so that the marine economy, resources, and environmental systems can develop a high coupling level and realize coordinated development.

2. In order to improve the marine ecological environment and enhance the carrying capacity of the marine ecosystem, statistical analysis of marine ecological pollution should be further strengthened. On this basis, we should strengthen legislation, enhance penalties for the main bodies that damage the marine ecological environment, encourage the development of a green marine economy, strengthen institutional innovation, and improve the supporting policies for the development of a green marine economy.

3. Change in the mode of economic growth and control of the consumption of resources and energy are necessary. China is in the industrialization stage at present. The driving force of economic development comes mainly from the development of infrastructure and some heavy industries. In this context, to change the mode of economic development and achieve sustainable development, we should, on the one hand, realize the optimal allocation of marine

resources and reduce energy consumption and the discharge of environmental pollutants. On the other hand, we need to optimize and upgrade the marine industrial structure and vigorously develop new industries with low pollution and high efficiency.

References

Allan, G., Hanley, N., Mcgregor, P., & Turner, K. (2007). The impact of increased efficiency in the industrial use of energy: A computable general equilibrium analysis for the United Kingdom. *Energy Economics, 29*(4), 779−798.

Chen, C., López, C. D., & Walker, B. L. E. (2014). A framework to assess the vulnerability of California commercial sea urchin fishermen to the impact of MPAs under climate change. *Geo Journal, 79*(6), 755−773.

Di, J., Porter, H., & Tracey, M. D. (2003). Linking economic and ecological models for a marine ecosystem. *Ecological Economics, 46*(3), 367−385.

Diao, X. D., Zeng, S. X., & Tam, C. M. (2009). EKC analysis for studying economic growth and environmental quality: A case study in China. *Journal of Cleaner Production, 17*, 541−548.

Field, J. G. (2003). The gulf of guinea large marine ecosystem: Environmental forcing and sustainable development of marine resources. *Journal of Experimental Marine Biology and Ecology, 296*(1), 128−130.

Gogoberidze, G. (2012). Tools for comprehensive estimate of coastal region marine economy potential and its use for coastal planning. *Journal of Coastal Conservation, 16*(3), 251−260.

Grimaud, A., & Rouge, L. (2008). Environment, directed technical change and economic policy. *Environmental and Resource Economics, 41*(10), 439−463.

Grossman, G. M., & Krueger, A. B. (1991). Environmental impacts of a North American free trade agreement. *Social Science Electronic Publishing, 8*(2), 223−250.

Halpern, B. S., Longo, C., Hardy, D., Mcleod, K. L., & Samhouri, J. F. (2012). An index to assess the health and benefits of the global ocean. *Nature, 488*(7413), 615−620.

Jonathan, S., & Paul, J. (2002). Technologies and their influence on future UK marine resource development and management. *Marine Policy, 26*(4), 231−241.

Li, X. M., Min, M., & Tan, C. H. (2004). The functional assessment of agricultural ecosystems in Hubei Province. *China Ecological Modelling, 187*(2), 352−360.

Liem, T., Tran, C., Gregory, K., Robert, V., Elizabeth, R. S., Kurt, H., & Riitters, J. W. (2002). Fuzzy decision analysis for integrated environmental vulnerability assessment of the mid-Atlantic region. *Environmental Management, 29*(6), 845−859.

Maddison, D. (2006). Environmental Kuznets curves: A spatial econometric approach. *Journal of Environmental Economics & Management, 51*(2), 218−230.

Managi, S., Opaluch, J. J., Jin, D., & Grigalunas, T. A. (2005). Technological change and petroleum exploration in the Gulf of Mexico. *Energy Policy, 33*(5), 619−632.

Manuel, B., William, W. L., Merino, G., & Perry, R. I. (2010). Modelling the potential impacts of climate change and human activities on the sustainability of marine resources. *Current Opinion in Environmental Sustainability, 2*(5-6), 326−333.

Oliverira, C., & Antunes, C. H. (2011). A multi-objective multi-sectoral economy−energy−environment model: Application to Portugal. *Energy, 36*(5), 2856−2866.

Orubu, C. O., & Omotor, D. G. (2011). Environmental quality and economic growth: Searching for environmental Kuznets curves for air and water pollutants in Africa. *Energy Policy, 39*(7), 4178−4188.

Qin, X. H., Sun, C. Z., & Wang, Z. Y. (2014). Marine economy sustainable development assessment of cities in the Bohai Sea Ring area from the perspective of metabolic cycle. *Resources Science, 36*(12), 2647—2656.

Read, P., & Field, T. (2003). Management of environmental impacts of marine aquaculture in Europe. *Aquaculture, 226*(1-4), 139—163.

Reuveny, R. (2002). Economic growth, environmental scarcity, and conflict. *Global Environmental Politics, 2*(1), 83—110.

Saboori, B., Sulaiman, J., & Mohd, S. (2012). Economic growth and CO_2 emissions in Malaysia: A cointegration analysis of the Environmental Kuznets curve. *Energy Policy, 51*(4), 184—191.

Salvador, E. P., José, L. P., & Mariana, C. G. (2007). Modeling population dynamics and economic growth as competing species: An application to CO_2 global emissions. *Ecological Economics, 65*(3), 602—615.

Samonte, T., White, A. T., Tercero, M. A., & John, D. (2007). Economic valuation of coastal and marine resources: Bohol marine triangle, Philippines. *Coastal Management, 35*(2-3), 319—338.

Slesser, M., & Hounam, I. (1984). *Carrying capacity assessment.* Rome: Report to UNESCO & FAO.

Stern, D. I., & Common, M. S. (2001). Is there an environmental Kuznets curve for sulfur? *Journal of Environmental Economics & Management, 41*(2), 162—178.

Sun, B. L., & Wang, A. M. (2012). The quantitative assessment of the system coordination of the marine economy—resources—environment in Zhejiang Province. *Forum on Science and Technology in China, 02*, 95—101.

Tiba, S., & Omri, A. (2017). Literature survey on the relationships between energy, environment and economic growth. *Renewable and Sustainable Energy Reviews, 69*, 1129—1146.

Vefie, L. (1996). *The Penguin dictionary of Physics.* (pp. 92—93). Beijing: Foreign language Press.

Wang, S. Y., Liu, J. S., & Yang, C. J. (2008). Eco-environmental vulnerability evaluation in the Yellow River Basin, China. *Pedosphere, 18*(2), 171—182.

Yasuhiro, S., & Kazuhiro, Y. (2005). Population concentration, urbanization and demographic transition. *Journal of Urban Economics, 58*(1), 45—61.

Zhang, P. Y., Su, F., Li, H., & Sang, Q. (2008). Coordination degree of urban population, economy, space, and environment in Shenyang since 1990. *China Population, Resources and Environment, 18*(2), 115—119.

CHAPTER EIGHT

Evaluation of marine resource management levels in China

8.1 Introduction

With the progress of science and technology, the exploitation of marine resources has received unprecedented attention, and the marine economy has developed very rapidly. However, China's ocean resource management mode was decentralized only 5 years ago. According to the provisions of relevant laws and regulations, maritime administrative organs are scattered in four primary departments of the State Council: the State Oceanic Administration, the Maritime Safety Administration, China Fisheries Law Enforcement, and the Ministry of Environmental Protection. Over a period of time, the functions of these marine resource management agencies overlapped. Although the relevant provincial departments are responsible for integrated marine resource management, it is hard to realize integration due to the lack of a higher authority. Similarly, the lack of a cross-sectoral coordination system in planning and management often results in the untimely management of marine resources (Shi, Hutchinson, Yu, & Xu, 2001). Although there are governments that attempted to achieve integration, sectoral self-interests and regional egoism clearly overthrew these coordination systems (Chou, 2005). To change the laxity of marine management, the Chinese government restructured the State Oceanic Administration in 2013, integrating the former State Oceanic Administration, Marine Monitoring, Coastal Guard of the Ministry of Public Security, Fisheries Supervision of the Ministry of Agriculture, and the General Administration of Customs. After the reorganization, maritime administration was transferred to the reorganized State Oceanic Administration. The State Council stipulates that the restructured State Oceanic Administration shall be administered by the Ministry of Land and Resources and its main tasks are to draft marine development plans, implement marine rights, manage marine utilization, and protect the marine environment. The reorganized State Oceanic Administration implements maritime rights in the name of the China Coast Guard and receives the operational guidance of the Ministry of Public Security. Since the reorganization, the State Oceanic Administration has strengthened the supervision and enforcement of marine management. Meanwhile, China has established the Ocean Commission, a special institution for studying the national marine development strategy and coordinating major marine affairs. The work of the Ocean Commission is undertaken by China's State Oceanic Administration. Hitherto, China's marine resource management has gradually changed from decentralization to centralization. The promulgation of China's Sea Laws and Regulations started relatively late. Since the 1980s, a series of marine laws and policies have been gradually formed in the legal system of marine resource management. There are more than 10 laws related to marine management, including the Environmental Protection Law (1989, amended in 2014), Soil and Water Conservation Law (1991 and 2010), Marine Environmental Protection Law (1999, amended in 2014), and Administration on Marine Utilization Management (2001). Other policies include the China Ocean Agenda for the 21st Century (1996), Focus and Distribution of Marine Industry Development (for 2011−2020) (2010), and the Twelfth Outline of the Five-Year Plan for National Economic and Social Development (for 2011−2015). However, China's marine legal system is far from perfect. For example, without the basic law of the sea, the existing law of the sea is poorly operable.

To summarize, compared to other countries, China's development and utilization of marine resources are relatively late, and the management system, in all aspects, is not perfect. According to the Announcement on the Use and Management of Sea Areas 2005−2015, issued by China's State Oceanic Administration, China's newly added sea area was 25,687.22 km^2, and the increasing use of sea areas throughout the years has exacerbated the environment of coastal areas year after year. Therefore to ensure the rapid and healthy development of the marine industry, marine resource management has become a problem that cannot be ignored.

8.2 Literature review

The management of natural resources has been the focus of scholars' attention. As a type of the natural resources of the largest size and strong development potentials, the ocean has attracted scholars' increasing attention. Kim (2012) divided the worldwide marine resource management system into five parts: interministerial committees or committees, management of subordinate departments at ministerial level, management of the subordinate organs of the ministerial and ministerial committees, and ministerial departments and ministerial committees. In 1999 Zhong discussed the sustainable development of natural resources in China, summarized the contributions of and challenges pertaining to natural resources, and proposed a series of suggestions on the planning and management of natural resources. Based on the idea of sustainable economic development, Zhao and Jing (2006) established a mathematical model for the effective allocation of natural resources and proposed research methods for natural resource management and suggested a series of principles of natural resource management for sustainable economic development. Dong (2006) constructed an evaluation system from a policy perspective, assessed the situation of marine resource management in Korea, and explored the opportunities and constraints of the Korean marine resource management policies. Carneiro (2007) evaluated the management of marine resources in Portugal, emphasizing the need for concrete measures to integrate the strategies and objectives of national marine resource management. Cao and Wong (2007) assessed the management plans and progress of coordinated legislation, institutional arrangements, public participation, capacity-building, and scientific research (mainly coastal planning and functional zoning) in Chinese coastal areas and studied recent socioeconomic and environmental problems in Chinese coastal areas. Song and Tsai (2007) presented recommendations for improving environmental management, utilization and distribution of natural resources, assessment of sustainable natural resource management, pollution prevention and treatment, and improvement of policies related to natural resources. Prokopy (2008) demonstrated the inadequacy of existing ethical norms for collaborative natural resource management by taking the case of management in three river basins in the central and western regions of China as examples. Although most researchers are supervised by the Institutional Review Board, there are still many decisions that researchers can make in the course of their research, which may unintentionally harm the individuals and the environment under study. Researchers focus on ethical considerations related to collecting data and publishing research results. Some feasible suggestions have been proposed, to guide research cooperation in natural resource management. Chang, Allen, Dawson, and Madsen (2012) used a social network method to explore the role of networks in natural resource management and proposed a method to identify the structure of regional networks. This method can quantitatively analyze natural resources and help managers formulate natural resource management plans. In addition, Wakita and Yagi (2013) assessed the development guidelines for Japan's integrated coastal zone management plan and analyzed the reasons for the poor implementation of the guidelines in 2000. Ye, Chou, Yang, Yang, and Du (2014) and Kong, Xue, and Mao (2015) analyzed the relationships among the marine management, coastal environment, and marine economy of Quanzhou, established 32 evaluation indexes suitable for Chinese coastal areas, and used the principal component analysis method as a weighting method to evaluate these indexes comprehensively. According to the evaluation results, the key performance index of marine management was determined, and specific suggestions were offered, to serve as a reference for future decision-making of marine management. Stern and Coleman (2015) attempted to decompose the concept of trust into components and reorganize trust theory in a robust and practical way to achieve collaborative natural resource management. They described four forms of trust related to the cooperative management of natural resources: disposal trust, rationality trust, kinship trust, and procedure trust. By describing the different forms of trust, its causes, and the potential consequences for collaborative management of natural resources, the study provided a consistent and useful vocabulary and framework for researchers and practitioners to use in the workforce aspect of natural resource management. Kenchington (2016) analyzed the evolution of the Australian regional marine management and provided suggestions for its future development. Xu, Dong, Teng, and Zhang (2018), using the ecosystem model, applied a multispatial model to evaluate the intensity of the marine development of Chinese coastal areas. The results showed that the intensity exhibited large spatial differences. Yan, Wang, and Xia (2018) analyzed the target orientation and challenges of the natural resource reform. By applying literature research and a comprehensive analysis, they contended that the target orientation of natural resource management reform is to achieve five unifications, namely, "unified investigation and evaluation, unified power registration, unified multi-conformity and use control, unified asset management, and unified examination and supervision." Existing challenges include the insufficient theoretical reserve on natural resource management, shortage of scientific guidance on development and utilization ethics, long-standing contradictions and conflicts between development and protection, and the to-be-tested effectiveness of the new management system. In view of these challenges, the reform of natural resource management should emphasize sustainable utilization, fair efficiency, and effective market construction to effectively reshape the new pattern of natural resource management. In the new era, the reform of natural resource management should be based on ownership and improve the natural resource market system, deem planning as the primary task for building a territorial spatial planning system, and promote the control of territorial and spatial utilization throughout the space of national land. Further, the reform should strengthen the comprehensive management of the natural community through renovation and improve the legal system of natural resource management through laws and regulations.

Based on the above literature, it can be seen that the management of marine resources has also attracted the attention of scholars. However, most of the existing studies are qualitative. Quantitative studies on the management level of marine resources in Chinese coastal areas are rare. Therefore in this chapter appropriate evaluation methods are selected to quantitatively evaluate the management level of marine resources in Chinese coastal areas and to compare the advantages and disadvantages of these areas to propose feasible suggestions for the development of regional marine resource management.

8.3 Evaluation index system and method

8.3.1 Evaluation index system

Crarnes, Cooper, Harrald, Karwan, and Wallace (1976) proposed that marine resource management is a method by which all kinds of activities (resource acquisition, platform establishment, etc.) occurring at sea and environmental quality are considered as a whole, which is to be utilized to obtain the greatest benefit for a country as regards sustainable development. However, this definition is too broad. Lv and Wang (2007), based on the research of many scholars, further improved the definition of marine resource management as follows: the management behaviors of sea-related departments and their personnel involved in sea-related activities by themselves and the management behaviors of social organizations.

To achieve a more objective and accurate empirical evaluation and to truly reflect the condition of the evaluated objects, the following principles should be followed for setting up and constructing an index system:

1. The principle of scientificity requires that indexes have certain scientific connotations, can be accurately defined and measured, have clear calculation methods and scopes, and can accurately, comprehensively, and truly reflect the condition of the evaluated objects.
2. The relevance principle requires that indexes have a certain relevance with the significance of the evaluated objects, leading to more accurate evaluation results.
3. The principle of comparability requires that indexes can be evaluated and compared in a unified way, regardless of the time, region, unit, or individual studied.
4. The principle of feasibility requires that indexes have certain information contents and can be operated and analyzed smoothly and that the data corresponding to the indexes can be obtained; that is to say, the indexes should be operable.

To ensure the accuracy and validity of the evaluation, in most cases, the multiindex comprehensive evaluation method is used. Marine resource management covers a wide range of aspects, including science and technology, environment, and administration. Therefore based on existing research this chapter divides the evaluation index system of marine resource management level into three parts: marine scientific and technological management capacity, marine environmental management capacity, and marine administrative management capacity. These are the first-level indexes; please refer to Table 8.1 for details.

8.3.2 Evaluation method

The comprehensive evaluation system of the level of marine resource management is composed of many different indexes. There are differences among the indexes in different regions. Therefore the role and importance of each index in the entire evaluation system are different. Current evaluation methods can be divided into two main categories. One is the objective weighting method, which determines the weight vector through the information in the statistical data itself (the correlations among indexes or the differences among indexes). The other is the subjective weighting method, which obtains the source information by consulting experts and determines the weight vector according to the method of comprehensive

Table 8.1 Evaluation indexes for marine resource management.

Evaluation indexes of marine resource management level	First-level index	Second-level index
	Scientific and technological management	X1: Number of sci-tech subjects in marine scientific and technological research institutions
		X2: Internal expenditure of R&D funds for marine sci-tech research institutions (CNY1000s)
		X3: Number of R&D subjects in marine sci-tech research institutions
		X4: Number of marine scientific and technological research institutions
		X5: Number of sea-related sci-tech employees
	Environmental management	X6: Number of pollution control projects
		X7: Quantity of general industrial solid waste disposal
		X8: Quantity of comprehensive utilization of general industrial solid waste
	Administrative management	X9: Number of coastal observatories
		X10: Number of certificates of right to use sea area issued
		X11: Proportion of authorized sea area

consultation score. The nuisance attribute model measures the attribute membership degree of an element by using an interval with a small change in its length, which is more accurate than a single numerical value. The model can effectively measure the attribute stages of the evaluated objects but has certain subjectivity in determining the weights of the indexes. In this chapter, to avoid deviating and unreasonable research results, it is proposed to combine the entropy weight method and expert scoring method to determine the weights, thus avoiding the defects of too many subjective influences.

8.3.2.1 Entropy weight method
The concept of entropy was proposed by German physicist Clausius in 1850 and was used to reflect the uniformity of energy distribution in space. The value of entropy is proportional to the uniformity of the energy distribution. The calculations in the entropy method are relatively

simple and there is no need to make assumptions about the distribution states of the data. Moreover, through calculations, it can be seen that the correlation among each index in this chapter is not high and accurate results cannot be obtained by using principal component analysis or factor analysis. Therefore this study uses the entropy method for empowerment.

In information theory, information entropy is defined as

$$\sum_i p(x_i) = 1 \tag{8.1}$$

where $p(x_i) \in [0, 1]$. Information entropy is used to represent the degree of variation of the indexes and to evaluate them comprehensively. Assume that there are m evaluation objects and n evaluation indexes used to form the original index data matrix $X = (x_{ij})_{m \times n}$. For a certain index x_i, the greater the gap between it and index X_{ij}, the greater the amount of information the index provides, the greater the role it plays in the comprehensive evaluation, the smaller the corresponding information entropy, and the greater the weight assigned to the index. Otherwise, this weight is smaller. If the values of the index are all equal, the index will not play a role in the comprehensive evaluation.

Let $p_{ij} = X_{ij}$ $(i = 1, 2, \ldots m)$ denote the standardized value of the ith index in the jth area and let n be the year of evaluation. Then, p_{ij} can be expressed as

Positive index:

$$p_{ij} = \frac{X_{ij} - \min_{1 \leq j \leq n} X_{ij}}{\max_{1 \leq j \leq n} X_{ij} - \min_{1 \leq j \leq n} X_{ij}} \tag{8.2}$$

Negative index:

$$p_{ij} = \frac{\max_{1 \leq j \leq n} X_{ij} - X_{ij}}{\max_{1 \leq j \leq n} X_{ij} - \min_{1 \leq j \leq n} X_{ij}} \tag{8.3}$$

The right hand side of formula (8.2) represents the relative distance between the gap of a certain index value to the minimum index value and the gap of the maximum index value and the minimum index value. The larger the difference, the larger the value after normalization. Formulas (8.3) and (8.2) have the same economic meaning.

Formula (8.4) is used to calculate the entropy of each index.

$$f_{ij} = \frac{p_{ij}}{\sum_{j=1}^{n} p_{ij}} \quad (i = 1, 2, \ldots, m; \ j = 1, 2, \ldots, n) \tag{8.4}$$

where $\sum_{j=1}^{n} p_{ij}$ represents the sum of the data of all the evaluated areas of the ith index and f_{ij} represents the feature proportion of the jth city under the ith index. If $f_{ij} = 0$, then $\lim_{f_{ij}=0} \ln f_{ij} = 0$; E_i is the entropy value of the ith index.

Let e_i be the entropy weight of the ith index, then:

$$e_i = \frac{1 - E_i}{m - \sum_{i}^{m} E_i}, \quad i = 1, 2, \ldots, m \tag{8.5}$$

where $\sum_{i=1}^{m} e_i = 1$.

8.3.2.2 Expert scoring method

The expert scoring method involves selecting appropriate evaluation indexes, to be given to experts, based on the practical significance of patent quality indexes, and providing the weight range of patent quality indexes, which are assigned by the scoring method as follows: Each expert marks each column of patent quality weights and assigns the final weights for each index. All experts discuss with each other whether a score is reasonable or not. If there is any objection, the score is regraded until found satisfactory. The average weight of each index is obtained after collecting all the scoring tables. The average weights obtained are recorded as the final weights, for which $\sum_{i=1}^{m} w_i = 1$.

Given the weights e_i and w_i, obtained by the entropy weight method and the expert scoring method, respectively, assume that the corresponding comprehensive weight of the ith index X_i is h_i. This weight is given by

$$h_i = \frac{w_i \times e_i}{\sum_{i=1}^{m} w_i \times e_i} \tag{8.6}$$

and $\sum_{i=1}^{m} h_i = 1$.

8.3.2.3 Attribute measurement

Let U denote the space set of an evaluated object, which contains the definition, u, of each element, the specific index corresponding to element u is defined as m, the attribute value of the index is defined as X_i $(i = 1, 2, \ldots, m)$, and the corresponding evaluation set for element u is defined as V_z $(z = 1, 2, \ldots, k)$ and indicates the quality grade or evaluation

Table 8.2 Grade classification of a given index.

Grade ($z = 1,2,\ldots,K$)	Index ($i = 1,2,\ldots,m$)			
	X_1	X_2	$\ldots$	X_m
V_1	$a_{10}-a_{11}$	$a_{20}-a_{21}$	$\ldots$	$a_{m0}-a_{m1}$
V_2	$a_{11}-a_{12}$	$a_{21}-a_{22}$	$\ldots$	$a_{m1}-a_{m2}$
$\ldots$	$\ldots$	$\ldots$	$\ldots$	$\ldots$
V_k	$a_{1k-1}-a_{1k}$	$a_{2k-1}-a_{2k}$	$\ldots$	$a_{mk-1}-a_{mk}$

Note: Classification of the grades for a given index, which can be obtained from the experts' opinions. The grade for a given index is determined based on the data by classifying indexes into several grades.

type of the index. Generally, the measured values are usually numerical. Table 8.2 provides the grades corresponding to the measurement values of a given index, in which the measured value of the ith index of element u is denoted by t_i.

According to Table 8.2, for the ith index X_i of element u, the attribute measurement function of the measured value t_i is denoted by (t), where z represents the grade z. There are two cases of a values: when a_{iz} is positive, $a_{i0} < a_{i1} < \cdots < a_{ik}$; when a_{iz} is negative, $a_{i0} > a_{i1} > \cdots > a_{ik}$. The corresponding single index attribute measurement functions are as follows:

1. When $a_{i0} < a_{i1} < \cdots < a_{ik}$ ($i = 1,2,\ldots,m;\ z = 1,2,\ldots,k-1$),

$$\mu_{ui1}(t) = \begin{cases} 1 & t < a_{i1} - d_{i1} \\ \dfrac{|t - a_{i1} - d_{i1}|}{2d_{i1}} & a_{i1} - d_{i1} \leq t \leq a_{i1} + d_{i1} \\ 0 & a_{i1} + d_{i1} < t \end{cases} \quad (8.7)$$

$$\mu_{uiz}(t) = \begin{cases} 0 & t < a_{iz-1} - d_{iz-1} \\ \dfrac{|t - a_{iz-1} + d_{iz-1}|}{2d_{iz-1}} & a_{iz-1} - d_{iz-1} \leq t \leq a_{iz-1} + d_{iz-1} \\ 1 & a_{iz-1} + d_{iz-1} < t < a_{iz} - d_{iz} \\ \dfrac{|t - a_{iz} - d_{iz}|}{2d_{iz}} & a_{iz} - d_{iz} \leq t \leq a_{iz} + d_{iz} \\ 0 & a_{iz} + d_{iz} < t \end{cases} \quad (8.8)$$

$$\mu_{uik}(t) = \begin{cases} 1 & a_{ik-1} + d_{ik-1} < t \\ \dfrac{|t - a_{ik-1} + d_{ik-1}|}{2d_{ik-1}} & a_{ik-1} - d_{ik-1} \leq t \leq a_{ik-1} + d_{ik-1} \\ 0 & t < a_{ik-1} - d_{ik-1} \end{cases} \quad (8.9)$$

where $d_{iz} = \min(|b_{iz} - a_{iz}|, |b_{iz+1} - a_{iz}|, z = 1, 2, \ldots, k-1)$; $b_{iz} = ((a_{iz-1} + a_{iz})/2),\ z = 1, 2, \ldots, k$.

2. When $a_{i0} > a_{i1} > \cdots > a_{ik}$ ($i = 1, 2, \ldots, m;\ z = 1, 2, \ldots, k-1$),

$$\mu_{ui1}(t) = \begin{cases} 1 & a_{i1} + d_{i1} < t \\ \dfrac{|t - a_{i1} + d_{i1}|}{2d_{i1}} & a_{i1} - d_{i1} \leq t \leq a_{i1} + d_{i1} \\ 0 & t < a_{i1} - d_{i1} \end{cases} \quad (8.10)$$

$$\mu_{uiz}(t) = \begin{cases} 0 & t < a_{iz} - d_{iz} \\ \dfrac{|t - a_{iz} + d_{iz}|}{2d_{iz}} & a_{iz} - d_{iz} \leq t \leq a_{iz} + d_{iz} \\ 1 & a_{iz} - d_{iz} < t < a_{iz-1} + d_{iz+1} \\ \dfrac{|t - a_{iz-1} - d_{iz-1}|}{2d_{iz-1}} & a_{iz-1} - d_{iz-1} \leq t \leq a_{iz-1} + d_{iz-1} \\ 0 & a_{iz-1} + d_{iz-1} < t \end{cases} \quad (8.11)$$

$$\mu_{uik}(t) = \begin{cases} 1 & t < a_{ik-1} - d_{ik-1} \\ \dfrac{|t - a_{ik-1} + d_{ik-1}|}{2d_{ik-1}} & a_{ik-1} - d_{ik-1} \leq t \leq a_{ik-1} + d_{ik-1} \\ 0 & a_{ik-1} + d_{ik-1} < t \end{cases} \quad (8.12)$$

where $d_{iz} = \min(|b_{iz} - a_{iz}|, |b_{iz+1} - a_{iz}|, z = 1, 2, \ldots, k - 1)$; $b_{iz} = ((a_{iz-1} + a_{iz})/2), \ z = 1, 2, \ldots, k.$

By introducing the original data and hierarchical data into the above attribute measurement functions, the attribute measure μ_{uiz} of each second-level index can be obtained. By introducing the weight h_i and attribute μ_{uiz} of each index into the following formula, the comprehensive attribute measurement, μ_{ui}, of each index can be obtained and can be used to make a final evaluation of the management level of marine resources.

$$\mu_{ui} = \sum_{i=1}^{m} h_i \mu_{uiz} \quad (8.13)$$

8.4 Evaluation and analysis of marine resource management level

8.4.1 Data source

According to the provisions of the *China Ocean Statistical Yearbook*, coastal areas refer to areas with coastlines (continental coastline and island coastline), which can be divided into coastal provinces, autonomous regions, and municipalities directly under the Central Government according to administrative regions. As of 2019 China has nine coastal provinces (Hebei, Liaoning, Shandong, Jiangsu, Zhejiang, Fujian, Guangdong, Hainan, and Taiwan), one autonomous region (Guangxi Zhuang Autonomous Region), and two municipalities directly under the Central Government (Tianjin and Shanghai). Based on the evaluation index system of marine resource management level constructed above, 11 coastal areas are taken as the research objects, the latest available data (2015) are selected, and the data of each index are obtained by consulting the *China Ocean Statistics Yearbook* and government bulletins, as shown in Table 8.3.

Formulas (8.2) and (8.3) are used to standardize the data on the various indexes. See Table 8.4.

8.4.2 Weight construction

According to the 2015 data and the opinions of relevant experts, the data of each index area were classified into three levels: poor (V_1), medium (V_2), and excellent (V_3), as shown in Tables 8.5−8.7.

Using the classification tables of the levels of marine resource management in Chinese coastal areas in 2015, the attribute measurement functions of these levels can be obtained. The results are shown in Table 8.8 (taking the number of sci-tech subjects in marine sci-tech research institutions as an example; results for other indexes are listed in the appendix).

Considering Tables 8.5−8.7 together with the original data on the level of marine resource management in Chinese coastal areas in 2015, we can obtain the attribute measurement of each index of this level, as shown in Table 8.9.

The objective weight coefficients are calculated based on the entropy method, and then, the subjective weight coefficients of each index are computed by the expert scoring method. Finally, through weighing, the comprehensive weight coefficients are obtained, as shown in Table 8.10.

Through the above analysis, the index attribute measurements and corresponding weights of marine resource management levels in Chinese coastal areas are finally obtained and then used in formula (8.7) to obtain the comprehensive attribute measurements of the 2015 marine resource management level in the 11 aforementioned areas.

Table 8.3 Original data of marine resource management evaluation in Chinese coastal areas in 2015.

Area	Number of sci-tech subjects in marine sci-tech research institutions	Internal expenditure of R&D funds for marine sci-tech research institutions (CNY1000s)	Number of R&D subjects in marine sci-tech research institutions	Number of marine sci-tech research institutions	Number of sea-related sci-tech employees	Number of pollution control projects	Disposal of general industrial solid waste (tons)	Comprehensive utilization of general industrial solid waste (tons)	Number of coastal observatories	Number of issued certificates of right to use sea area	Proportion of confirmed sea area (%)
Tianjin	790	895,039	442	16	2808	25	21.5	1524	25	37	0.4029
Hebei	77	91,209	61	5	552	17	14,729.1	19,900	37	558	2.6611
Liaoning	685	1023,051	478	22	3151	30	8067.3	10,028.9	170	1098	2.7595
Shanghai	1166	2307,215	772	15	3989	80	72.2	1796.2	148	5	0.1624
Jiangsu	2187	830,532	1558	10	3356	225	407.4	10,207	94	218	0.8924
Zhejiang	645	556,497	300	21	2028	328	205.3	4263.2	158	247	0.0964
Fujian	618	366,234	530	14	1189	151	1157.4	3784.3	227	288	0.1781
Shandong	1637	1801,784	1290	22	4108	102	737.2	18,308.6	151	680	2.3666
Guangdong	2653	1778,965	2261	26	5434	194	438.8	5102.7	213	161	0.2593
Guangxi	125	84,997	107	11	1225	43	546.2	4387.7	41	276	0.8698
Hainan	25	3001	13	3	265	3	41.6	268.1	89	33	0.000847

Table 8.4 Standardized data of marine resource management evaluation in Chinese coastal areas in 2015.

Area	Number of sci-tech subjects in marine sci-tech research institutions	Internal expenditure of R&D funds for marine sci-tech research institutions (CNY1000s)	Number of R&D subjects in marine sci-tech research institutions	Number of marine sci-tech research institutions	Number of sea-related sci-tech employees	Number of pollution control projects	Disposal of general industrial solid waste (tons)	Comprehensive utilization of general industrial solid waste (tons)	Number of coastal observatories	Number of issued certificates of right to use sea area	Proportion of confirmed sea area (%)
Tianjin	0.07447	0.09191	0.05658	0.09697	0.09991	0.02087	0.48539	0.15536	0.01848	0.01027	0.00403
Hebei	0.00726	0.00937	0.00781	0.03030	0.01964	0.01419	0.00030	0.00505	0.02735	0.15496	0.02661
Liaoning	0.06457	0.10505	0.06119	0.13333	0.11212	0.02504	0.00106	0.01929	0.12565	0.30492	0.02759
Shanghai	0.10992	0.23692	0.09882	0.09091	0.14193	0.06678	0.21308	0.19433	0.10939	0.00139	0.00162
Jiangsu	0.20617	0.08528	0.19944	0.06061	0.11941	0.18781	0.03187	0.02886	0.06948	0.06054	0.00892
Zhejiang	0.06080	0.05714	0.03840	0.12727	0.07216	0.27379	0.06860	0.07495	0.11678	0.06859	0.00096
Fujian	0.05826	0.03761	0.06784	0.08485	0.04231	0.12604	0.01331	0.09238	0.16778	0.07998	0.00178
Shandong	0.15432	0.18502	0.16513	0.13333	0.14617	0.08514	0.03848	0.03515	0.11160	0.18884	0.02367
Guangdong	0.25009	0.18267	0.28943	0.15758	0.19335	0.16194	0.07815	0.15248	0.15743	0.04471	0.00259
Guangxi	0.01178	0.00873	0.01370	0.06667	0.04359	0.03589	0.00491	0.01387	0.03030	0.07665	0.00870
Hainan	0.00236	0.00031	0.00166	0.01818	0.00943	0.00250	0.06484	0.22828	0.06578	0.00916	0.00001

Table 8.5 Classification of administrative levels for marine resource management evaluation in Chinese coastal areas in 2015.

Grade	Number of sci-tech subjects in marine sci-tech research institutions	Internal expenditure of R&D funds for marine sci-tech research institutions (CNY1000s)	Number of R&D subjects in marine sci-tech research institutions	Number of marine sci-tech research institutions	Number of sea-related sci-tech employees
V_1	0−0.05	0−0.03	0−0.02	0−0.06	0−0.04
V_2	0.05−0.15	0.03−0.12	0.02−0.10	0.065−0.13	0.04−0.12
V_3	0.15−0.25	0.12−0.24	0.10−0.30	0.13−0.25	0.12−0.20

Table 8.6 Assessment of marine management level in Chinese coastal areas in 2015.

Grade	Number of pollution control projects	Disposal of general industrial solid waste (tons)	Comprehensive utilization of general industrial solid waste (tons)
V_1	0−0.02	0−0.005	0−0.02
V_2	0.02−0.15	0.005−0.07	0.02−0.15
V_3	0.15−0.30	0.07−0.50	0.15−0.25

Table 8.7 Classification of administrative levels for evaluation of marine resources management level in Chinese coastal areas in 2015.

Grade	Number of coastal observatories	Number of issued certificates of right to use sea area	Proportion of confirmed sea area (%)
V_1	0−0.05	0−0.04	0−0.001
V_2	0.05−0.12	0.04−0.15	0.001−0.02
V_3	0.12−0.25	0.15−0.30	0.02−0.03

8.4.3 Analysis of results

As regards the level of science and technology management (Fig. 8.1), Hainan, Hebei, and Guangxi receive a poor grade; Fujian, Zhejiang, Liaoning, and Tianjin earn a medium grade; and Shanghai, Shandong, Jiangsu, and Guangdong attain an excellent grade. These results are closely related to the degree of development of each area. The scientific and technological support of regional scientific research institutions and institutions of higher learning is an important aid in scientific and technological management. Shanghai, Shandong, Jiangsu, and Guangdong, the

Table 8.8 Attribute measurement function of marine resource management level in Chinese coastal areas in 2015.

Index	Poor	Medium	Excellent
Number of sci-tech subjects in marine sci-tech research institutions	$\mu_{x11} = \begin{cases} 1 & t < 0.025 \\ 1 - \dfrac{|t-0.075|}{0.05} & 0.025 \le t \le 0.075 \\ 0 & 0.075 < t \end{cases}$	$\mu_{x12} = \begin{cases} 0 & t < 0.025 \\ \dfrac{|t-0.025|}{0.05} & 0.025 \le t \le 0.075 \\ 1 & 0.075 < t < 0.1 \\ 1 - \dfrac{|t-0.2|}{0.1} & 0.1 \le t \le 0.2 \\ 0 & 0.2 < t \end{cases}$	$\mu_{x13} = \begin{cases} 0 & t < 0.1 \\ 1 - \dfrac{|t-0.1|}{0.1} & 0.1 \le t \le 0.2 \\ & 0.2 < t \end{cases}$
Internal expenditure of R&D funds for marine sci-tech research institutions (CNY1000s)	$\mu_{x21} = \begin{cases} 1 & t < 0.015 \\ 1 - \dfrac{|t-0.075|}{0.05} & 0.015 \le t \le 0.045 \\ 0 & 0.045 < t \end{cases}$	$\mu_{x22} = \begin{cases} 0 & t < 0.015 \\ \dfrac{|t-0.015|}{0.03} & 0.015 \le t \le 0.045 \\ 1 & 0.045 < t < 0.075 \\ 1 - \dfrac{|t-0.165|}{0.09} & 0.075 \le t \le 0.165 \\ 0 & 0.165 < t \end{cases}$	$\mu_{x23} = \begin{cases} 0 & t < 0.075 \\ 1 - \dfrac{|t-0.045|}{0.09} & 0.075 \le t \le 0.165 \\ & 0.165 < t \end{cases}$
Number of R&D subjects in marine sci-tech research institutions	$\mu_{x31} = \begin{cases} 1 & t < 0.01 \\ 1 - \dfrac{|t-0.03|}{0.05} & 0.01 \le t \le 0.03 \\ 0 & 0.03 < t \end{cases}$	$\mu_{x32} = \begin{cases} 0 & t < 0.01 \\ \dfrac{|t-0.01|}{0.02} & 0.01 \le t \le 0.03 \\ 1 & 0.03 < t < 0.06 \\ 1 - \dfrac{|t-0.14|}{0.08} & 0.06 \le t \le 0.14 \\ 0 & 0.14 < t \end{cases}$	$\mu_{x33} = \begin{cases} 0 & t < 0.06 \\ 1 - \dfrac{|t-0.06|}{0.08} & 0.06 \le t \le 0.14 \\ & 0.14 < t \end{cases}$
Number of marine sci-tech research institutions	$\mu_{x41} = \begin{cases} 1 & t < 0.03 \\ 1 - \dfrac{|t-0.09|}{0.06} & 0.03 \le t \le 0.09 \\ 0 & 0.09 < t \end{cases}$	$\mu_{x42} = \begin{cases} 0 & t < 0.03 \\ \dfrac{|t-0.03|}{0.05} & 0.03 \le t \le 0.09 \\ 1 & 0.09 < t < 0.0975 \\ 1 - \dfrac{|t-0.2|}{0.1} & 0.0975 \le t \le 0.1625 \\ 0 & 0.1625 < t \end{cases}$	$\mu_{x43} = \begin{cases} 0 & t < 0.0975 \\ 1 - \dfrac{|t-0.0975|}{0.065} & 0.0975 \le t \le 0.1625 \\ & 0.1625 < t \end{cases}$

Number of sea-related sci-tech employees

$$\mu_{x51} = \begin{cases} 1 & t<0.02 \\[4pt] \dfrac{|t-0.06|}{0.04} & 0.02\le t\le 0.06 \\[4pt] 0 & 0.06<t \end{cases}$$

$$\mu_{x52} = \begin{cases} 0 & t<0.02 \\[4pt] \dfrac{|t-0.02|}{0.04} & 0.02\le t\le 0.06 \\[4pt] 1 & 0.06<t<0.08 \\[4pt] \dfrac{|t-0.16|}{0.08} & 0.08\le t\le 0.16 \end{cases}$$

$$\mu_{x53} = \begin{cases} 1 & 0.16<t \\[4pt] \dfrac{|t-0.08|}{0.08} & 0.08\le t\le 0.16 \\[4pt] 0 & t<0.08 \end{cases}$$

Number of pollution control projects

$$\mu_{x61} = \begin{cases} 1 & t<0.01 \\[4pt] \dfrac{|t-0.03|}{0.02} & 0.01\le t\le 0.03 \\[4pt] 0 & 0.03<t \end{cases}$$

$$\mu_{x62} = \begin{cases} 0 & t<0.01 \\[4pt] \dfrac{|t-0.01|}{0.02} & 0.01\le t\le 0.03 \\[4pt] 1 & 0.03<t<0.095 \\[4pt] \dfrac{|t-0.215|}{0.13} & 0.095\le t\le 0.215 \\[4pt] 0 & 0.215<t \end{cases}$$

$$\mu_{x63} = \begin{cases} 1 & 0.215<t \\[4pt] \dfrac{|t-0.095|}{0.13} & 0.095\le t\le 0.215 \\[4pt] 0 & t<0.095 \end{cases}$$

Disposal of general industrial solid waste (tons)

$$\mu_{x71} = \begin{cases} 1 & t<0.0025 \\[4pt] \dfrac{|t-0.0075|}{0.005} & 0.0025\le t\le 0.0075 \\[4pt] 0 & 0.0075<t \end{cases}$$

$$\mu_{x72} = \begin{cases} 0 & t<0.0025 \\[4pt] \dfrac{|t-0.0025|}{0.005} & 0.0025\le t\le 0.0075 \\[4pt] 1 & 0.0075<t<0.0375 \\[4pt] \dfrac{|t-0.1025|}{0.065} & 0.0375\le t\le 0.1025 \\[4pt] 0 & 0.1025<t \end{cases}$$

$$\mu_{x73} = \begin{cases} 1 & 0.1025<t \\[4pt] \dfrac{|t-0.0375|}{0.065} & 0.0375\le t\le 0.1025 \\[4pt] 0 & t<0.0375 \end{cases}$$

Comprehensive utilization of general industrial solid waste (tons)

$$\mu_{x81} = \begin{cases} 1 & t<0.01 \\[4pt] \dfrac{|t-0.03|}{0.02} & 0.01\le t\le 0.03 \\[4pt] 0 & 0.03<t \end{cases}$$

$$\mu_{x82} = \begin{cases} 0 & t<0.01 \\[4pt] \dfrac{|t-0.01|}{0.02} & 0.01\le t\le 0.03 \\[4pt] 1 & 0.03<t<0.1 \\[4pt] \dfrac{|t-0.2|}{0.1} & 0.1\le t\le 0.2 \\[4pt] 0 & 0.2<t \end{cases}$$

$$\mu_{x83} = \begin{cases} 1 & 0.2<t \\[4pt] \dfrac{|t-0.1|}{0.1} & 0.1\le t\le 0.2 \\[4pt] 0 & t<0.2 \end{cases}$$

(Continued)

Table 8.8 (Continued)

Index	Poor	Medium	Excellent

Number of coastal observatories

$$\mu_{x91} = \begin{cases} 1 & t < 0.025 \\ \dfrac{|t - 0.075|}{0.05} & 0.025 \le t \le 0.075 \\ 0 & 0.075 < t \end{cases}$$

$$\mu_{x92} = \begin{cases} 0 & t < 0.025 \\ \dfrac{|t - 0.025|}{0.05} & 0.025 \le t \le 0.075 \\ 1 & 0.075 < t < 0.085 \\ \dfrac{|t - 0.155|}{0.07} & 0.085 \le t \le 0.155 \\ 0 & 0.155 < t \end{cases}$$

$$\mu_{x93} = \begin{cases} \dfrac{|t - 0.085|}{0.07} & 0.085 \le t \le 0.155 \\ 0 & t < 0.085 \end{cases} \quad 0.155 < t$$

Number of issued certificates of right to use the sea area

$$\mu_{x101} = \begin{cases} \dfrac{1}{|t - 0.06|} & t < 0.02 \\ 0.04 & 0.02 \le t \le 0.06 \\ 0 & 0.06 < t \end{cases}$$

$$\mu_{x102} = \begin{cases} 0 & t < 0.02 \\ \dfrac{|t - 0.02|}{0.04} & 0.02 \le t \le 0.06 \\ 1 & 0.06 < t < 0.095 \\ \dfrac{|t - 0.205|}{0.11} & 0.095 \le t \le 0.205 \\ 0 & 0.205 < t \end{cases}$$

$$\mu_{x103} = \begin{cases} \dfrac{|t - 0.095|}{0.11} & 0.095 \le t \le 0.205 \\ 0 & t < 0.095 \end{cases} \quad 0.205 < t$$

Proportion of confirmed sea area (%)

$$\mu_{x111} = \begin{cases} \dfrac{1}{|t - 0.0015|} & t < 0.0005 \\ 0.001 & 0.0005 \le t \le 0.0015 \\ 0 & 0.0015 < t \end{cases}$$

$$\mu_{x112} = \begin{cases} 0 & t < 0.0005 \\ \dfrac{|t - 0.0005|}{0.001} & 0.0005 \le t \le 0.0015 \\ 1 & 0.0015 < t < 0.015 \\ \dfrac{|t - 0.025|}{0.01} & 0.015 \le t \le 0.025 \\ 0 & 0.025 < t \end{cases}$$

$$\mu_{x113} = \begin{cases} \dfrac{|t - 0.015|}{0.01} & 0.015 \le t \le 0.025 \\ 0 & t < 0.015 \end{cases} \quad 0.025 < t$$

Table 8.9 Attribute measurement of level of marine resource management in Chinese coastal areas in 2015.

Area	Scientific and technological management index	Environmental management index	Administrative management index
Hainan	(1,0,0) (1,0,0) (1,0,0) (1,0,0) (1,0,0)	(1,0,0) (0,0.579,0.421) (0,0,1)	(0.184,0.816,0) (1,0,0) (1,0,0)
Hebei	(1,0,0) (1,0,0) (1,0,0) (0.995,0.005,0) (1,0,0)	(0.79,0.21,0) (1,0,0) (1,0,0)	(0.953,0.047,0) (0,0.455,0.545) (0,0.161,0.839)
Guangxi	(1,0,0) (1,0,0) (0.815,0.185,0) (0.389,0.611,0) (0.41,0.59,0)	(0,1,0) (0.518,0.482,0) (0.807,0.193,0)	(0.894,0.106,0) (0,1,0) (0,1,0)
Fujian	(0.334,0.666,0) (0.246,0.754,0) (0,0.902, 0.098) (0.086,0.914,0) (0.442,0.558,0)	(0,0.684,0.316) (0,1,0) (0,1,0)	(0,0,1) (0,1,0) (0.281,0.719,0)
Zhejiang	(0.284,0.716,0) (0,1,0) (0,1,0) (0,0.542,0.458) (0,1,0)	(0,0,1) (0,0.521,0.479) (0,1,0)	(0,0.546,0.454) (0,1,0) (0.536,0.464,0)
Liaoning	(0.209,0.791,0) (0,0.666,0.334) (0,0.985,0.015) (0,0.449,0.551) (0,0.599,0.401)	(0.248,0.752,0) (1,0,0) (0.536,0.464,0)	(0,0.419,0.581) (0,0,1) (0,0,1)
Tianjin	(0,1,0) (0,0.812,0.188) (0,1,0) (0,1,0) (0,0.751,0.249)	(0.457,0.543,0) (0,0,1) (0,0.446,0.554)	(1,0,0) (1,0,0) (0,1,0)
Shanghai	(0,0.9,0.1) (0,0,1) (0,0.515,0.485) (0,1,0) (0,0.226,0.774)	(0,1,0) (0,0,1) (0,0.057,0.943)	(0,0.652,0.348) (1,0,0) (0,1,0)
Shandong	(0,0.457,0.543) (0,0,1) (0,0,1) (0,0.449,0.551) (0,0.173,0.827)	(0,1,0) (0,0.984,0.016) (0,1,0)	(0,0.62,0.38) (0,0.147,0.853) (0,0.133,0.867)
Jiangsu	(0,0,1) (0,0.886,0.114) (0,0,1) (0.49,0.51,0) (0,0.507,0.493)	(0,0.209,0.791) (0,1,0) (0,0.057,0.943)	(0.11,0.89,0) (0,1,0) (0,1,0)
Guangdong	(0,0,1) (0,0,1) (0,0,1) (0,0.076, 0.924) (0,0,1)	(0,0.408,0.592) (0,0.374,0.626) (0,0.475,0.525)	(0,0,1) (0.382,0.618,0) (0,1,0)

Note: The attribute measurement value of each index is expressed as (x, y, z), where x, y, and z represent the proportion of poor, medium, and excellent grades, respectively.

Table 8.10 Weights of marine resource management level indexes in Chinese coastal areas in 2015.

Index name	Subjective	Objective	Final weight
Number of sci-tech subjects in marine sci-tech research institutions	0.0541	0.0863	0.0443
Internal expenditure of R&D funds for marine sci-tech research institutions (CNY1000)	0.0946	0.0825	0.0741
Number of R&D subjects in marine sci-tech research institutions	0.0676	0.0967	0.0620
Number of marine sci-tech research institutions	0.1081	0.0459	0.0471
Number of sea-related sci-tech employees	0.1081	0.0588	0.0603
Number of pollution control projects	0.1216	0.0927	0.1070
Disposal of general industrial solid waste (tons)	0.1081	0.1734	0.1780
Comprehensive utilization of general industrial solid waste (tons)	0.0946	0.0892	0.0801
Number of coastal observatories	0.0946	0.0615	0.0553
Number of issued certificate of right to use the sea area	0.0676	0.0956	0.0613
The proportion of confirmed sea area (%)	0.0811	0.1175	0.0904

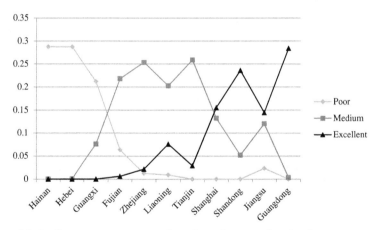

Figure 8.1 Attribute measurements of marine science and technology management levels in Chinese coastal areas.

developed cities in China, are in the leading positions in all aspects and have many scientific research institutes. Good governmental policies also support the implementation of science and technology and promote the development of regional science and technology management.

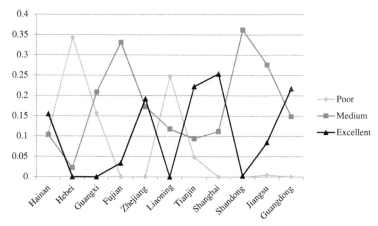

Figure 8.2 Attribute measurements of marine environmental management levels in Chinese coastal areas.

As regards the level of environmental management (Fig. 8.2), the marine environmental management of Hebei and Liaoning is poor. The marine development around Bohai Sea is rapid, but the environmental management has not kept up with the development speed. Hence, many systems need to be improved and the methods and intensity of environmental improvement need to be strengthened. The environmental management level of Hainan, Zhejiang, Tianjin, Shanghai, and Guangdong is excellent. Tianjin, Shanghai, and Guangdong attach great importance to environmental management and have strong means for enforcing environmental management. Hainan and Zhejiang exhibit less marine development, which makes the management of marine environment easier. Guangxi, Fujian, and Shandong are at the medium level. Similarly to Hainan and Zhejiang, the marine development of Guangxi and Fujian is relatively low, making the management easier.

In terms of administrative level (Fig. 8.3), Hebei, Liaoning, and Shandong attain an excellent grade. Although these three provinces exhibit low grades in science and technology and environmental management, they still have a dominant position in terms of law enforcement and sea observation and control. The administrative management of Tianjin and Hainan needs to be strengthened.

The comprehensive attribute rating of marine resource management (Fig. 8.4) is obtained by considering Figs. 8.1–8.3 together. Shanghai and Guangdong, at the first level, are the growth poles of the Pearl River Delta and Yangtze River Delta regions and they all have relatively perfect

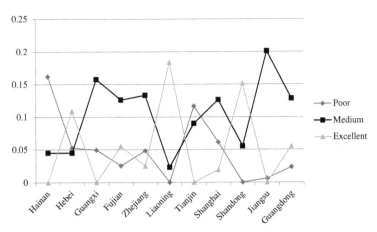

Figure 8.3 Attribute measurements of marine administrative management levels in Chinese coastal areas.

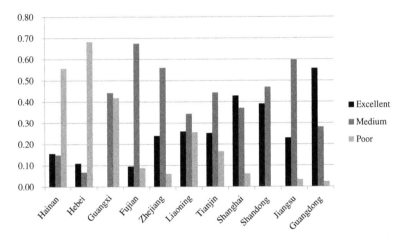

Figure 8.4 Comprehensive attribute ratings of marine resources management level in Chinese coastal areas.
Note: Each index in the graph has three columns, representing an excellent, medium, and poor rating. The highest of the three columns is the evaluation grade of the area.

administrative management models. Meanwhile, there are many scientific research institutes, such as Shanghai Ocean University and Guangdong Ocean University, which provide good science and technology management support. Their governments attach great importance to environmental governance issues and, accordingly, their environmental management ability is strong. The seven areas at the medium level also have corresponding rules and regulations for marine management, but most provinces have few available sea areas. It can be noted that Shandong does not contribute to the proportion of areas at the poor level, as its marine resource management capability is outstanding. Shandong has more than 3000 km of coastline and many ports and is a traditional coastal province. However, due to some disadvantages in administrative management, it cannot achieve an excellent level. Tianjin is the center of the Bohai economic circle and a core port city. Tianjin's Binhai New Area has become a national new area vigorously supported by the country since its inclusion in the national development strategy in 2005. The economic development of the new area has greatly promoted the management level of marine resources in Tianjin and the improvement of this level also promotes the development of the new area. The marine development area of the Guangxi Zhuang Autonomous Region is small and mainly used for open aquaculture with a weak ability for scientific and technological management and environmental management. Similarly, the coastal areas of Jiangsu mainly develop aquaculture and agriculture, but unlike Guangxi, Jiangsu has a good reserve of scientific research. As a leading province in education in China, Jiangsu province has excellent scientific research strength and can provide technological support for marine resource management. It is not difficult to see from the attribute measurement table that Fujian has a good level of administrative management, a perfect management system of fishery headquarters and other departments, and strict law enforcement. However, the inadequate capacity for environmental management and scientific and technological management is also a common problem in some coastal areas of southern China. Zhejiang's marine resource development intensity is low, but the marine economy is developed and its management mode is worth studying and being referenced. The management grades for the Hainan and Hubei provinces are relatively low; most are at the poor level in the attribute measurement of each index, which is related to the fact that they have less available sea area and to their weak economy. The development and utilization of marine resources and marine scientific research are not sufficient in Hainan and Hubei and there are no sufficient scientific research institutes for technical support. From the government bulletins, we can observe that Hainan and Hubei still have defects in system formulation, system implementation, law enforcement, and inspection. These factors have affected their management grades of marine resources.

8.5 Conclusion and recommendations

This chapter constructs the management level system of marine resources in Chinese coastal areas and evaluates the indexes of the system through a comprehensive evaluation method, an entropy weight method, and an expert evaluation method. The empirical results show that the 11 coastal areas in China can be divided into 3 grade levels: poor, medium, and excellent. Guangdong and Shanghai are at the forefront of the country and belong to the excellent level. Guangxi, Fujian, Zhejiang, Jiangsu, Tianjin, and Shandong are at the medium level. Hainan, Hebei, and Liaoning are at the relatively poor level. Comparing the three levels, we can find that the developed provinces and cities pay much more attention to marine resource management than the economically weaker areas. The marine resource management levels of the 11 areas are closely related to their economic development levels.

Based on the results obtained, the following recommendations can be offered:

1. Strengthen the concept of marine resource management and optimize the industrial structure

    The scale of China's marine resource development is gradually expanding. Therefore attention should be paid to protecting the original marine ecosystem during marine resource development. The results of level evaluation show that the levels of environmental management in most areas are medium or poor. Attention to environmental management should not only exist in developed areas and the concept of marine ecology should be advocated in all coastal areas. It is recommended that certain restrictions should be imposed on bays and estuaries with excessive development intensity to achieve the goal of adjusting development intensity and improving the quality of the marine environment. Marine resource management departments should further standardize relevant regulations to promote the effective utilization of marine resources, fully exploit the regional advantages, optimize and adjust the marine industrial structure, and promote the sustainable development of the marine economy.

2. Increase investment in scientific research and enhance management ability

    Since China's reform and opening up, economic development has promoted the continuous progress of science and technology. How to use science and technology to drive the development of the marine industry and promote the management level of marine resources is an issue of marine development that cannot be ignored. Coastal areas should strengthen their investment in sea-related scientific research and enhance the regional exchange of relevant technologies. Further, technologically developed areas should help the development in technologically weaker areas. Appropriate point-to-point support measures should be taken to enhance scientific and technological strength and enact the scientific development, utilization, and management of the oceans.

3. Improve laws and regulations and strengthen law enforcement

The promulgation of sea-related laws and regulations in China started relatively late. Although a series of laws and regulations have been introduced after the implementation of centralized management by the central government, they are still far from perfect and the operability of the existing system is not enough. Therefore we should standardize the management regulations, strictly control and improve the standards of issuing certificates of the right to use sea areas, increase the enforcement of the law in all provinces, and effectively guarantee the development of the marine industry.

References

Cao, W., & Wong, M. H. (2007). Current status of coastal zone issues and management in China: A review. *Environment International, 33*(7), 985—992.

Carneiro, G. (2007). The parallel evolution of ocean and coastal management policies in Portugal. *Marine Policy, 31*(4), 421—433.

Crames, A., Cooper, W. W., Harrald, J., Karwan, K. R., & Wallace, W. A. (1976). A goal interval programming model for resource allocation in a marine environmental protection program. *Journal of Environmental Economics and Management, 3*(4), 347—362.

Chang, C. Y., Allen, C., Dawson, E., & Madsen, G. E. (2012). Network analysis as a method for understanding the dynamics of natural resource management in rural communities. *Society & Natural Resources, 25*(1), 203—208.

Chou, C. H. (2005). *A study on China's marine affairs strategy and organization structure*. Taipei: Coastal Guard Administration of Executive Yuan. (in Chinese).

Dong, O. C. (2006). Evaluation of the ocean governance system in Korea. *Marine Policy, 30*(5), 570—579.

Kenchington, R. (2016). *The evolution of marine conservation and marine protected areas in Australia. Big, bold and blue: Lessons from Australia's marine protected areas*. Clayton: CSIRO Publishing.

Kim, S. G. (2012). The impact of institutional arrangement on ocean governance: international trends and the case of Korea. *Ocean & Coastal Management, 64*, 47—55.

Kong, H., Xue, X., & Mao, Z. (2015). Towards integrated coastal governance with Chinese characteristics—A preliminary analysis of China's coastal and ocean governance with special reference to the ICM practice in Quanzhou. *Ocean & Coastal Management., 111*, 34—49.

Lv, J. H., & Wang, G. (2007). Marine administration: The orientation and construction of a public management perspective. *Ocean Development and Management*, *24*(6), 41−44.

Prokopy, L. S. (2008). Ethical concerns in researching collaborative natural resource management. *Society & Natural Resources*, *21*(3), 258−265.

Shi, C., Hutchinson, S. M., Yu, L., & Xu, S. (2001). Towards a sustainable coast: An integrated coastal zone management framework for Shanghai, People's Republic of China. *Ocean & Coastal Management*, *44*(5−6), 411−427.

Song, Y. H., & Tsai, C. L. (2007). *Japan marine policy development and strategy*. Policy suggestion.

Stern, M., & Coleman, K. (2015). The multidimensionality of trust: Applications in collaborative natural resource management. *Society & Natural Resources*, *28*(2), 117−132.

Wakita, K., & Yagi, N. (2013). Evaluating integrated coastal management planning policy in Japan: Why the guideline 2000 has not been implemented. *Ocean & Coastal Management*, *84*, 97−106.

Xu, W., Dong, Y. E., Teng, X., & Zhang, P. P. (2018). Evaluation of the development intensity of China's coastal area. *Ocean & Coastal Management*, *157*, 124−129.

Yan, J. M., Wang, X. L., & Xia, F. Z. (2018). Reshaping the new pattern of natural resource management: Target orientation, value orientation and strategic choice. *China Land Sciences*, *4*.

Ye, G., Chou, L. M., Yang, L., Yang, S., & Du, J. (2014). Evaluating the performance of integrated coastal management in Quanzhou, Fujian, China. *Ocean & Coastal Management*, *96*(96), 112−122.

Zhao, G. H., & Jing, S. T. (2006). Study on natural resources management for sustainable economic development in China. In: *Third international conference on management of technology*, Taiyuan, People's Republic of China.

Zhong, Z. (1999). Natural resources planning, management, and sustainable use in China. *Resources Policy*, *25*(4), 211−220.

Index

Printed in the United States
By Bookmasters